Assessing the Impact of U.S. Air Force National Security Space Launch Acquisition Decisions

An Independent Analysis of the Global Heavy Lift Launch Market

BONNIE L. TRIEZENBERG, COLBY PEYTON STEINER, GRANT JOHNSON,
JONATHAN CHAM, EDER SOUSA, MOON KIM, MARY KATE ADGIE

Prepared for the United States Air Force
Approved for public release; distribution unlimited

 PROJECT AIR FORCE

For more information on this publication, visit www.rand.org/t/RR4251

Library of Congress Cataloging-in-Publication Data is available for this publication.
ISBN: 978-1-9774-0399-5

Published by the RAND Corporation, Santa Monica, Calif.
© Copyright 2020 RAND Corporation
RAND® is a registered trademark.

Cover: Courtesy photo by United Launch Alliance.

Support RAND
Make a tax-deductible charitable contribution at
www.rand.org/giving/contribute

www.rand.org

Preface

The U.S. Air Force (USAF) today stands at a crossroads with respect to its space vehicle launch acquisition strategy. For years, it has relied on Atlas V and Delta IV rockets to launch heavy National Security Space (NSS) payloads—typically those weighing 5,000 pounds or more—for defense and intelligence clients. However, the Atlas V is powered by Russian engines. Under U.S. law, the USAF is prohibited after 2022 from purchasing rockets with engines designed or manufactured in Russia. Moreover, the Delta family will be retired from service over the next few years. Despite these retirements, the USAF is nevertheless required to sustain at least two viable launch vehicle providers for NSS payloads, and those payloads must be launched from U.S. soil. With 30-plus unprocured NSS launches forecasted for the 2022–2025 period and replacement launch vehicles still in development, the USAF's ability to fulfill vital NSS missions has come under substantial congressional scrutiny.

The USAF asked RAND Project AIR FORCE to independently analyze the heavy lift launch market to assess how decisions the USAF might make in the near term could affect domestic launch providers and the market in general. The research reported here was commissioned by the Air Force Space and Missile Systems Center and conducted within the Force Modernization and Employment Program of RAND Project AIR FORCE as part of a fiscal year 2019 project entitled *Space Launch Market Research and Assessment*. The intended audience for this report includes the U.S. Department of Defense, USAF leadership, and the U.S. Congress.

RAND Project AIR FORCE

RAND Project AIR FORCE (PAF), a division of the RAND Corporation, is the U.S. Air Force's federally funded research and development center for studies and analyses. PAF provides the Air Force with independent analyses of policy alternatives affecting the development, employment, combat readiness, and support of current and future air, space, and cyber forces. Research is conducted in four programs: Strategy and Doctrine; Force Modernization and Employment; Manpower, Personnel, and Training; and Resource Management. The research reported here was prepared under contract FA7014-16-D-1000.

Additional information about PAF is available on our website: www.rand.org/paf/

This report documents work originally shared with the U.S. Air Force on July 9, 2019.

Contents

Figures

Tables

Summary

Issue

For many years, the U.S Air Force (USAF) has relied on Atlas V and Delta IV rockets to launch heavy National Security Space (NSS) payloads—typically those weighing 5,000 pounds or more—for defense and intelligence clients. But that arrangement is soon to change. The Atlas V is powered by Russian engines and the Delta family is being retired from service in the next few years. Under U.S. law, the USAF is prohibited after 2022 from purchasing rockets with engines designed or manufactured in Russia. Moreover, the USAF is required to sustain at least two viable launch vehicle providers for NSS payloads, and those payloads must be launched from U.S. soil. With 30-plus unprocured NSS launches forecasted for the 2022–2025 period and replacement launch vehicles still in development, the USAF's ability to fulfill vital NSS missions has come under substantial congressional scrutiny.[1]

Research Questions

The USAF asked RAND Project AIR FORCE (PAF) to independently analyze the heavy lift launch market to assess how decisions it might make in the near term could affect domestic launch providers.[2] To conduct the research, PAF examined historical and projected levels of demand and supply in the global commercial and NSS launch markets to understand several issues:

- the number of U.S. launch service providers that the global heavy lift launch market can support

---

[1] This scrutiny focuses on USAF organizational ability, capacity, and planning, not the technical abilities of USAF personnel.

[2] We were tasked to identify costs, risks, and impacts of these decisions. To mitigate the possibility that cost identification would affect the Phase 2 source selection activities, we address costs at an abstract level only and focus primarily on risks and impacts.

- the impact of near-term acquisition decisions on the USAF's ability to (1) meet NSS launch demand using U.S. launch service providers and (2) sustain two or more U.S. launch service providers over the next ten years.

Risks Identified by Our Study

Specifically, we were asked to identify costs, risks, and effects to U.S. launch providers, the U.S. government, and the commercial markets more generally. Over the course of the study, we identified three primary risks that the USAF should consider when making near-term launch acquisition decisions.

- The national security risk of not having assured access to space in times of need because of a shortage of NSS-certified launch vehicles. Although this report identifies and quantifies a short-term issue because of insufficient supply, this risk might also arise if a certified provider is removed from the market temporarily because of a launch failure or permanently after a business failure or market consolidation. Therefore, this report explores the risk of market consolidation as a function of U.S. firms' business strategies. Mitigating this national security risk is our top priority in formulating our recommendations.
- The increased acquisition costs that the USAF could incur if the burden of maintaining two NSS launch providers cannot be shared with the commercial market. As we will discuss, this has been the situation for the past two decades, changing only recently with SpaceX's certification for NSS launch. Although near-term USAF decisions might shape this risk in ways that are beneficial to both the U.S. government and U.S. launch service providers, it is unlikely to be avoided altogether.
- The technical risk of any individual NSS launch failure. Although the USAF has an impressive record of NSS launch reliability, sustaining a near 100 percent success rate, there are legitimate concerns that the resources needed to maintain that record may be stretched thin if the USAF were to select three launch providers for near-term acquisitions.

Mitigating each of these risks to an acceptable level may require more resources than the U.S. government currently allocates to the acquisition of national security launch services and the sustainment of assured access to space. Because choices and balance among these risks are such critical issues, we requested clarification from Space and Missile Systems Center's Launch Enterprise regarding the current prioritization of these risks:

> The Launch Enterprise's top priority is mission success. In addition, the guidance from Air Force and DoD leadership, the White House Staff, and Congress was to urgently end the use of the Russian RD-180. Accordingly, when developing our

acquisition strategy, we prioritized ensuring mission success over the risk of only having a single provider for a short period of time (i.e., an Assured Access to Space Gap). Phase 2 involves transitioning our most critical payloads to new launch systems, ending the use of the Russian engine. We deliberately focus Phase 2 execution on mission assurance to sustain 100% mission success but will work to minimize the risk to Assured Access to Space to the maximum extent practicable.[3]

In light of the magnitude of the assured access risk we found in this study, we believe a larger policy conversation regarding resource allocation is needed. We encourage the U.S. Congress and the larger NSS community to engage in a meaningful dialog with the USAF regarding how to prioritize these launch-related risks within the larger context of all national security risks that the USAF must balance.

Conclusions

Global Commercial Heavy Lift Launch Market

- The annual number of launches worldwide grew to 71 in 2018 from 47 in 1998, but the commercial portion over which launch firms compete—*addressable share*—stayed steady at around 20 launches; today, the addressable share represents only 35 percent of the total market.
- SpaceX's Falcon 9 handles more than half of addressable launches today, with communications companies being the dominant customers.
- This market is likely to see only moderate growth over the next ten years, and the U.S. share of this market is likely to fall. The net result is that the addressable share of the launch market is unlikely to support more than one U.S. supplier of launch services focused on commercial heavy lift.

NSS Launch Market

- In many conceivable futures, the USAF's current acquisition plan is unlikely to provide sufficient near-term supply of NSS-certified launch vehicles.
- The USAF can lower the risk of insufficient near-term supply of NSS launch vehicles by exercising options under its Phase 1A contract or by negotiating contingency clauses in its Phase 2 launch vehicle contract, or by otherwise supporting three launch service providers through the 2022–2025 time frame.

---

[3] Email communication with author, February 21, 2020.

Recommendations

USAF Near-Term Strategy

- Continue to provide tailored support through 2023 to enable three U.S. launch service providers to continue in or enter the heavy lift launch market.
- Tailored support can be provided in many ways and this is not necessarily a recommendation to select three launch service providers for the NSS Phase 2 contract. In fact, depending on a firm's strategic choices, selection as a NSS provider might adversely affect a firm's ability to compete in the heavy lift launch market.
- Supporting three U.S. launch service providers in the short term might increase the probability of global supplier consolidation but decrease the probability of additional foreign competition. Additionally, this strategy
  - provides time for U.S. firms to best position themselves in the launch markets.
  - allows market forces to determine which firms are strongest, and thus survive, and which ones should exit.

USAF Longer-Term Strategy

- Make prudent preparations for a future with only two U.S. NSS-certified heavy lift launch providers, at least one of which might have little support from the commercial marketplace.

Acknowledgments

We are deeply indebted to the Launch Enterprise team at Space and Missile Systems Center for facilitating our understanding of Space and Missile Systems Center's plans regarding the future of National Security Space launch. We would also like to thank the numerous unattributed men and women in the launch and space services community who shared their thoughts with us regarding the future of the commercial launch market—our work is incalculably richer for their insights. Additionally, we would like to thank Mike Kennedy, George Nacouzi, and Tom Light of the RAND Corporation who served as our internal "red team," keeping us on track and focused. Chad Ohlandt, Gary McLeod, and Brian Dolan provided independent review of our final work to ensure its objectivity and rigor. Finally, we would like to thank Jordan Higginson for keeping us organized, and Maria Vega, Gordon Lee and Dave Richardson for their invaluable help in editing and cross-checking our work.

Abbreviations

CDR	critical design review
CNSA	Chinese National Space Agency
COTS	Commercial Orbital Transportation Services
CRS	Commercial Resupply Services
DoD	U.S. Department of Defense
ERB	Engineering Review Board
ESA	European Space Agency
FAA	Federal Aviation Administration
FCC	Federal Communications Commission
FFRDC	Federally Funded Research and Development Center
FTC	Federal Trade Commission
GAO	Government Accountability Office
GEM	Graphite Epoxy Motor
GEO	geosynchronous Earth orbit
GPS	Global Positioning System
GSLV	Geosynchronous Satellite Launch Vehicle
GTO	Geosynchronous Transfer Orbit
HEO	highly elliptical orbit
ICBM	Intercontinental Ballistic Missile
ILS	International Launch Services

IRL	Innovation Readiness Level
ISRO	Indian Space Research Organization
ISS	International Space Station
ITAR	International Trafficking in Arms
JAXA	Japan Aerospace Exploration Agency
LEO	low Earth orbit
LSA	Launch Service Agreement
MEO	medium Earth orbit
NASA	National Aeronautics and Space Administration
NGIS	Northrop Grumman Innovation Systems
NRO	National Reconnaissance Office
NSS	National Security Space
OTA	Other Transactional Authority
PAF	Project AIR FORCE
PDR	preliminary design review
PSLV	Polar Satellite Launch Vehicle
RFP	request for proposal
ULA	United Launch Alliance
USAF	U.S. Air Force

CHAPTER ONE

Introduction

The U.S. Air Force (USAF) today stands at a crossroads with respect to its launch acquisition strategy. Currently, it has awarded launch service contracts for the Atlas V, Delta IV Heavy, Falcon 9, and Falcon Heavy to meet its anticipated demand for National Security Space (NSS) launches through 2021 and partially into 2022.[1] It has issued a request for proposal (RFP) to award just over 30 additional launches beginning in 2020 to cover launches planned in 2022 and beyond—this is designated as the Phase 2 procurement. Given the currently forecasted demand for NSS launches, these awards should carry the USAF through the 2026 time frame, at which time another round of awards (Phase 3) would begin.[2]

There is considerable controversy regarding this plan in Congress and elsewhere. As a result, the USAF asked RAND Project AIR FORCE (PAF) to independently analyze the heavy lift launch market to assess how decisions it might make in the near term could affect domestic launch providers. To conduct the research, we examined historical and projected levels of demand and supply in the global commercial and NSS launch markets to understand the following issues:

- the number of U.S. launch service providers that the global heavy lift launch market can support
- the impact of near-term acquisition decisions on the USAF's ability to (1) meet NSS launch demand using U.S. launch service providers and (2) sustain two or more U.S. launch service providers over the next ten years.[3]

---

[1] The USAF is the launch agent responsible for obtaining and overseeing launch services for NSS missions for defense and intelligence clients. NSS launch criteria is defined in Chapter Two of this report. During the study period of performance, options were exercised for two Delta VI Heavy's for launches in 2023 and 2024.

[2] Notionally, the Phase 3 procurement decisions would be made in 2024 to support 2026 launch dates. Awards are made at least two years in advance of need to accommodate the detailed engineering required when integrating payloads with launch vehicles.

[3] We considered three near-term USAF acquisition decisions: (1) a decision not to procure additional Atlas V launch vehicles before a 2022 congressionally mandated ban on purchasing rockets with engines designed or manufactured in Russia goes into effect; (2) a decision to award contracts to two instead of three U.S. launch service providers, which is a key aspect of the ongoing Phase 2 procurement plan for NSS launches, specifically; and

This report documents the results of our analysis.

Our analysis and recommendations draw primarily from four categories of data described in Appendix A: (1) literature review, (2) historical launch data, (3) actor-specific launch data, and (4) interviews with subject-matter experts and industry stakeholders. Using these data sources, we developed a forecast for the future heavy lift launch markets and constructed vignettes to illuminate strategies U.S. firms could use to position themselves in the marketplace. Finally, we used a series of Monte Carlo simulations to determine how near-term USAF acquisition decisions might play out over a range of possible futures. Our methodology is described in Appendix B.

Launch Capability and National Security

Access to Space Is a Key Component to National Security

U.S. national security depends on space, which has made the assured access to heavy lift launch capabilities a long-standing policy concern. The United States is not alone in the desire to bolster national security through access to and uses of space. Although the United States, Russia, China, and members of the European Union are the primary spacefaring nations, the number of nations investing in domestic launch capabilities is growing. India and Japan's domestic launch capabilities are well established. Israel, Iran, North Korea, South Korea, and New Zealand have demonstrated capabilities. Argentina, Brazil, and Indonesia are all seeking to develop domestic capabilities.[4]

Services supplied from space provide essential infrastructure for economic and military power projection. Space systems provide precision timing that is critical to the accuracy of bank transactions and weapon systems. Instantaneous, worldwide communication across space links enables such diverse uses as live television transmissions of breaking news, ship-to-shore transactions, and nuclear weapons' command and control. Space-based remote sensing systems monitor and measure economic activity, predict the weather, and collect intelligence that aids in over-the-horizon targeting of adversary militaries. Should any of those systems be damaged or destroyed in a time of international conflict, a domestic launch capability would be essential to recover many basic services on which both citizens and militaries rely.

Definition of Heavy Lift Launch as Used in This Report

The term *heavy lift launch* in this report is defined by the class of launch vehicles capable of taking the heaviest payloads to orbit. Over time, the average mass of satel-

---

(3) a decision to terminate research and development agreements with providers that were not awarded a launch service contract under Phase 2.

[4]   Federal Aviation Administration (FAA), *The Annual Compendium of Commercial Space Transportation: 2018*, Washington, D.C., 2018.

lites above 2,000 kg has become heavier, while the average mass of satellites less than 2,000 kg is falling rapidly. Hence, the heavy lift launch vehicles of 1998 have become the medium lift launch vehicles of 2018. Furthermore, distinctions between heavy, medium, and small lift launch vehicles have become *more* distinct over time and distinctions between heavy lift and interplanetary launch vehicles have become *less* distinct. Chapter Two discusses the foreign and domestic launch vehicles that are currently competitive in or seeking entry into this heavy lift launch market.

Definition of Participants in the Launch Ecosystem as Used in This Report

Figure 1.1 illustrates the players in the launch market ecosystem. Launch service providers supply launch vehicles and services in response to demand from a space service provider. The space service provider, in turn, is responding to demand from end users (e.g., civilian, military, private firms, or governments) for communications, imagery, weather, timing (e.g., Global Positioning System [GPS]), and other space-enabled services. Payloads carried into space on the launch vehicle provide these space-enabled services to the end users. Usually the payload is a satellite or group of satellites.

When "the provider" is used in this report without a modifier, the term refers to the "launch service provider." In all other references to a provider, we have used a modifier to identify what is being provided (e.g., "communications service provider" or "payload provider").

Figure 1.1
The Launch Market Ecosystem

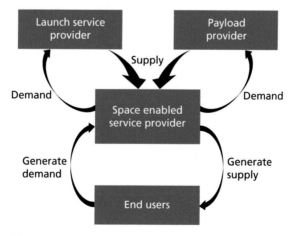

SOURCE: RAND analysis.

U.S. Law and NSS Launch Acquisitions

Under U.S. law, the USAF is the agent charged with procuring heavy lift services for defense and intelligence payloads. The actual provider of the space-enabled service might be one of the armed services or the National Reconnaissance Office (NRO). As the procurement agency, the USAF is charged with sustaining at least two viable launch vehicle providers for NSS payloads at all times.[5] The USAF is constrained by two additional conditions: NSS launches must be from U.S. soil and the USAF is prohibited from buying launch services after 2022 for vehicles that use engines designed or manufactured in Russia. The latter prohibition was first imposed in 2014 and subsequently revised; the governing statute today is contained in Section 1602 of the National Defense Authorization Act of Fiscal Year 2017.[6]

For many years, the USAF relied on the Atlas V and the Delta IV families of vehicles (both tracing their origins to intercontinental ballistic missile (ICBM) technologies in the late 1950s) to fulfill their obligation to sustain access to two launch vehicle families.[7] However, the Atlas V is powered by Russian engines and the Delta IV family is being retired from service by its provider, the United Launch Alliance (ULA). More recently, the USAF has certified SpaceX's Falcon 9 for NSS launches and SpaceX's Falcon Heavy is nearing completion of NSS nonrecurring certification

---

[5]   10 U.S. Code §2273 reads as follows:

> (a) Policy—It is the policy of the United States for the President to undertake actions appropriate to ensure, to the maximum extent practicable, that the United States has the capabilities necessary to launch and insert United States national security payloads into space whenever such payloads are needed in space.

> (b) Included Actions—The appropriate actions referred to in subsection (a) shall include, at a minimum, providing resources and policy guidance to sustain—

> (1) the availability of at least two space launch vehicles (or families of space launch vehicles) capable of delivering into space any payload designated by the Secretary of Defense or the Director of National Intelligence as a national security payload;

> (2) a robust space launch infrastructure and industrial base; and

> (3) the availability of rapid, responsive, and reliable space launches for National Security Space programs to—

> (A) improve the responsiveness and flexibility of a National Security Space system;

> (B) lower the costs of launching a National Security Space system; and

> (C) maintain risks of mission success at acceptable levels. (U.S. Code, Title 10, Section 2273, Policy Regarding Assured Access to Space: National Security Payloads, in effect as of January 7, 2011)

[6]   Public Law 114-328, National Defense Authorization Act for Fiscal Year 2017, Section 1602, (c)(2), Exception to the Prohibition on Contracting with Russian Suppliers of Rocket Engines for the Evolved Expendable Launch Vehicle Program, December 26, 2016.

[7]   "Atlas ICBM Chronology," webpage, last updated February 2006.

efforts.[8] The Falcon 9 was developed in the mid-2000s and matured with support from the National Aeronautics and Space Administration's (NASA) launch service provider program. The Falcon Heavy variant recently completed its third successful launch. Unless the USAF procures additional Atlas V's prior to the 2022 suspense date, the only available launch vehicles will be in the Falcon family.[9]

In an effort to increase the diversity of options available to them, the USAF has provided seed funding under other transactional authorities (OTAs) to help U.S. firms mature three new heavy lift launch vehicles and complete nonrecurring NSS certification efforts. However, none of these launch vehicles has a forecasted first launch date before 2021.

Two Launch Markets: Global Commercial and NSS-Certified Launch

Our analysis recognizes two distinct (but related) heavy lift launch markets: the global commercial launch market and the NSS-certified launch market. The global market is truly international but, as we will discuss later, exists largely because nation states allow it to exist. Both supply and demand can be heavily affected by regulatory issues within each nation, such as the U.S. restriction that NSS payloads be launched by U.S. launch providers from U.S. soil. Demand in the global commercial market has been steady but cyclical, driven by the capital expenditure cycle of the underlying telecommunications service providers who dominate this market. There is considerable speculation regarding the future of this demand source because of disruptions in the telecommunications market and in space services more generally. Supply is largely driven by the decisions of individual service providers.

The second market is the NSS-certified launch market. U.S. national priorities, government budgets, and national space architectures determine the market's demand. Although this demand is reasonably well known over a five-year period, it tends to slip further out into the future. Supply in this market is limited to launch service providers certified to carry U.S. NSS payloads. The NSS-certified launch market is a monopsony—a market in which there is a single dominant buyer—and the U.S. government is the only buyer. Depending on the number of firms in the NSS market and

---

[8] Nonrecurring NSS launch certification is conducted once for new launch vehicle design and includes qualifying the design and processes to build the new launch vehicle and verifying test hardware and software meet qualification standards. Recurring NSS launch certification is conducted for every launch vehicle to evaluate the workmanship and integration of each flight system.

[9] As we will discuss later, the USAF does have unexercised options under the Phase 1A contract to a limited number of legacy launch vehicles.

their strategies for positioning themselves in the overall global launch markets, this can lead to market breakdown.[10]

The global commercial and NSS-certified launch markets follow different dynamics in that global commercial launch providers face a business chasm between the completion and competition phases as shown in Figure 1.2. In the NSS market, providers that complete NSS nonrecurring certification (which includes a specified number of successful launches) are immediately included in future NSS competitions. In the commercial marketplace, on the other hand, providers seeking to find a foothold face a variety of challenges. These challenges are called the "chasm" by venture capitalists, who developed the model of Innovation Readiness Levels (IRLs) used in Figure 1.2.[11] As noted in Figure 1.2, USAF decisions regarding NSS affect U.S. firms attempting to enter the commercial market at every stage of the IRL. In the early prototyping phase, government investments help firms mature their offerings. A withdrawal of funds during the completion phase, as contemplated by the USAF for competitors not selected for Phase 2 contracts, could induce firms to abandon their efforts to enter the commercial market and provides an opportunity for foreign competitors to enter or expand their offerings. Selection as an NSS provider can help or hurt a firm's ability to cross the chasm, depending on the firm's strategic choices regarding how they want to position themselves in the market. We will return to this discussion later in the report as it is critical to understanding the impact USAF decisions may have on their ability to sustain two or more launch service providers for NSS launch. Finally, in

---

[10]   See Appendix C for more information on the ULA merger. There is significant evidence suggesting that the U.S. government's monopsony power might have contributed to firms offering to sell at a loss, leading to the eventual merger and loss of competitive supply.

[11]   Ming-Chang Lee, To Chang, and Wen-Tien Chang Chien, "An Approach for Developing Concept of Innovation Readiness Levels," *International Journal of Managing Information Technology*, Vol. 3, No. 2, May 2011. The levels are defined as:

> **Concept:** Basic scientific principles of the innovation have been observed and reported, and the critical functions and/or characteristics have been confirmed through experiments (equivalent to TRL 1-3).

> **Components:** Components have been developed and validated, and a prototype has been developed to demonstrate the technology (equivalent to TRL 4-6).

> **Completion:** Technological development has been completed and the complete system functionality has proven in the field (equivalent to TRL 7-9).

> **Chasm:** Refers to the challenges and difficulties that a new entrant or product may encounter when first introduced to the market (early stage).

> **Competition:** This is the mature phase of the market, when it has reached a state of equilibrium marked by the absence of significant growth or innovation. The main mission in this phase is to maintain and enhance the position of the product within the market and to cope with competition.

> **Changeover/Closedown:** These are the two options in the declining stage of the market. (Lee and Chien, 2011).

Figure 1.2
NSS-Certified and Global Commercial Market Timelines and NSS Potential Impacts on the Global Commercial Market

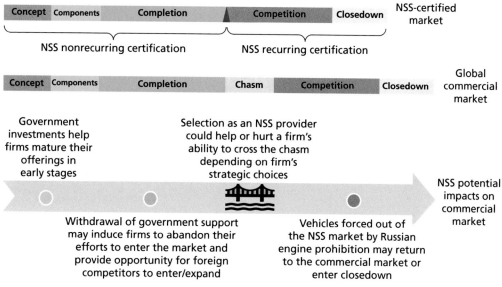

SOURCE: RAND analysis of market interactions using the IRL concept.

the competition phase, U.S. government decisions to remove launch vehicles from the NSS market—as with its decision to prohibit vehicles with engines designed or manufactured in Russia—may force providers of those vehicles into other markets or cause them to enter the closedown phase.[12]

National Investments in Improving Heavy Lift Launch Capabilities

Given national interests in launch capability, domestic launch service providers receive support from their respective governments in a wide variety of ways. In some nations, this support is provided by direct government ownership of the launch provider; in other nations, support is provided via direct government funding of research and development. Still others provide support via payments to cover operating costs, government-provided infrastructure, government funded launch contracts, guarantees of a fixed number of launches, or guaranteed indemnity insurance.[13] Figure 1.3 shows our

---

[12] Vehicles are removed from the NSS market only if they fail certifications or are otherwise prohibited by U.S. law. Selection as a provider under a given USAF launch contract is independent of NSS certification status of the vehicles.

[13] U.S. Government Accountability Office, *Evolved Expendable Launch Vehicle: DOD Is Assessing Data on Worldwide Launch Market to Inform New Acquisition Strategy*, Washington, D.C., July 22, 2016.

Figure 1.3
Government Investments in Improved Heavy Lift Launch Capabilities, 2005–2018

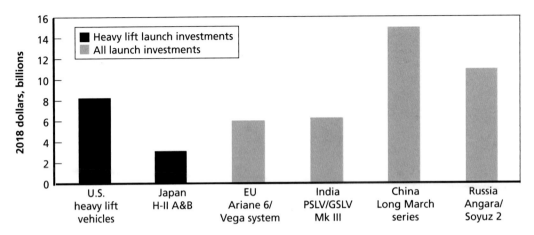

SOURCE: RAND analysis.
NOTES: Japan and U.S. investments show only those made in heavy lift launch capabilities, the subject of this report. European Union, Indian, and Russian investments are overstated in comparison because of the inclusion of investments in small and medium lift launch vehicles that we were unable to isolate out of the analysis. China's investments are likely highly overstated because of the inclusion of investments in small, medium, lunar, and interplanetary launch vehicles. EU = European Union. PSLV = Polar Satellite Launch Vehicle.

calculations of national investments in new or upgraded heavy lift launch capabilities made by the major spacefaring nations since 2005.[14] Foreign investments are likely overestimated because of our inability to properly isolate investments in the heavy lift launch market. Further details are provided in Appendix C.

National Investments in Launch Capability Affect the Global Commercial Launch Market

To understand how national investments in launch capability affect the global commercial launch market, we first need to understand how a capability deeply entwined with national security became a global commercial market. We also need to understand the relative size of the global launch market that is addressable by commercial firms.

Creation of the Global Commercial Launch Market

The commercial use of space has its roots in NASA's July 1963 launch of the world's first geo-synchronous communication satellite. Syncom 2, and the 1964 launch of

---

[14] To the extent possible, we have tried to isolate investments made in the heavy lift launch market that are the subject of this report. A breakdown of the U.S. investment in heavy lift launch is given later in this report when we discuss NSS launch.

Syncom 3, proved the concept that would lead to the first commercial communication satellite, Intelsat I (also known as Early Bird) in April 1965. These satellites weighed only 39 kg dry and 68 kg when fully fueled and were launched by a Delta D rocket into low Earth orbit (LEO) where an attached apogee kick motor was used to complete the transfer to geosynchronous Earth orbit (GEO).[15] These launches established the commercial satellite communications market and received worldwide public attention with Syncom 3's live television broadcast of the 1964 Tokyo Olympics. The "Live, via satellite!" motto defined the state of the art in communications for at least the next decade. Even today, demand for commercial heavy lift launch capabilities is driven largely by the telecommunications market, whether for telephony, television, radio, internet, or other uses.

U.S. manufacturers dominated the early satellite communications market. Naturally, they launched on U.S. domestic launch vehicles: Atlas, Delta, and Titan. These vehicles were originally developed during the Cold War in response to the perceived Soviet ICBM missile gap and were later converted from ICBM to launch vehicles to support U.S. efforts to put a man on the surface of the moon. In 1976, in an effort to reduce space launch cost, the United States funded the development of a reusable launch vehicle—the Space Transportation System, more commonly known as the space shuttle.[16] It was used extensively for commercial, civil, and U.S. Department of Defense (DoD) satellite launches from 1981 until the Challenger accident in 1986. For perspective, in 1985, four of the shuttle's nine missions (44 percent) carried commercial satellites to orbit.[17] U.S. policy regarding commercial launch in this period was

---

[15] Geosynchronous orbit is at a height above the Earth where a satellite appears stationary relative to a specific region on Earth. Therefore, satellites in this orbit are ideally situated to transmit and receive information from and to approximately 25 percent of the Earth within their field of view. These first satellites were positioned to provide communications across the Atlantic or Pacific oceans in an era when the capacity of intercontinental cables across oceans was quite limited. Satellite mass, orbits, and launch details can be found in the NASA Space Science Data Coordinated Archive (see NASA, "Syncom 2," Washington, D.C., NASA Space Science Data Coordinated Archive/COSPAR ID: 1963-031A, undated).

[16] Justifications for the space shuttle investment echo many of the themes being used today to justify additional U.S. government investment in space launch capability. In replying to the 1972 shuttle cost benefit report, NASA Administrator James Fletcher stated:

> The shuttle will provide quick and routine access to space and eliminate the constraints imposed by the present mode of space operations which is characterized by high risk, long lead times, and complex systems. [...] The low risk access to space possible with the shuttle will increase commercial interest in exploiting space in a wide variety of beneficial applications. (Comptroller General of the United States, *Cost Benefit Analysis Used in Support of the Space Shuttle Program: Report to Congress*, Washington, D.C.: U.S. Government Accountability Office, B-173777, June 2, 1972)

Although initial testing on a suborbital prototype began in 1976, the shuttle did not enter service until 1981 (NASA, undated).

[17] The other five launches supported a mix of civil and military users. NASA, "Kennedy Space Center Launch Archives: 1981–1986 Space Shuttle Launches," webpage, last updated February 24, 2008.

contradictory. As John Logsdon notes in his article regarding the history of the launch industry for the *Encyclopedia Britannica*:

> The U.S. had offered to turn over ownership and operation of existing expendable launch vehicles such as Delta, Atlas, and Titan to the private sector for commercial use; at the same time, it pursued an aggressive policy of marketing the space shuttle as a commercial launcher. The private sector could not compete with this government activity.[18]

After Challenger, however, shuttle usage for commercial launch was prohibited, forcing companies seeking to do business in space to find alternative sources for launch services.[19] Use of foreign launch providers was enabled through the granting of export licenses by the Department of State and, in the case of China, by presidential agreement.[20] The European Space Agency's Arianespace, the world's first commercial launch company (formed in 1980) rapidly gained market share, using its French Guiana spaceport to efficiently lift U.S. manufactured commercial satellites to space. U.S. commercial space companies also began to use China's Long March and Russia's Proton for transporting satellites to orbit.[21] The Atlas and Delta launch vehicles also saw a resurgence in use for commercial satellites once the shuttle was no longer competing in the commercial launch market. By 1998, a decade later, a robust global commercial space launch market had emerged with Arianespace as the dominant provider. Also in 1998, Congress passed the Commercial Space Act requiring the federal government to acquire space launch services from U.S. commercial providers as a commercial item, and the FAA's Commercial Space Transportation Office began to issue annual reports on the commercial space market.[22]

---

[18]   John M. Logsdon, "Launch Vehicle," *Encyclopedia Britannica*, March 13, 2019. Logsdon is a respected NASA historian and was the director of the Space Policy Institute at George Washington University from 1987 to 2008. In 2013, he was awarded the Frank J. Malina Astronautics Medal for outstanding contributions in space policy decisionmaking, space history, and education by the International Astronautical Federation.

[19]   For comparison, of the shuttle's eight launches in 1988, none carried a commercial satellite to space.

[20]   Export licensing of U.S. satellites is governed by International Trafficking in Arms Restrictions (ITAR) and the Arms Export Control Act.

[21]   Long March (China) became an option in September 1988 when President Ronald Reagan approved its use after the Challenger accident. Ten years later, export licensing for satellites was tightened after Space Systems/Loral and Hughes were accused of transferring militarily sensitive technology to China during launch accident investigations. Ryan Zelenio, "A Short History of Export Control Policy," *Space Review*, 2006.

[22]   The FAA's Commercial Space Transportation Office has published an annual list of global launches from 1994 to 2017. The 2018 list has been delayed because of the government shutdown in fiscal year 2019.

Understanding the Addressable Commercial Launch Market

This report seeks to understand the impact of U.S. national security launch acquisition decisions on U.S. launch providers who may wish to participate in the global commercial launch market. To do that, we need to define the portion of the market that is addressable by those firms. As noted above, the commercial launch market is highly regulated. The Outer Space Treaty holds nations responsible for all objects launched from their territory. Each spacefaring nation has its own law governing registration, ownership, and liability for objects launched into space from their territory. [23] Further complicating market dynamics, export laws in the United States have been used to deny the export of U.S. satellites to other nations for launch. U.S. export compliance was used as an effective lever to cut China out of the commercial launch market in the mid-1990s.[24] Perhaps as a result, Chinese built satellites rarely launch on commercial vehicles—China has effectively walled off its launch market from the rest of the world. In fact, a large number of launches, and not just those in China, are not addressable by commercial firms.

For the purposes of this report, we define the addressable commercial launch market by what it does not include, as follows:[25]

- **Launches that are subject to international agreements:** These include launches to the International Space Station (ISS), launches where government-owned entities package launch services with financing (e.g., China's Belt and Road Initiative), and launches negotiated through nation-to-nation agreements.
- **National security launches:** For those nations with domestic launch capability, we assume that the launch of satellites essential to national security will not be addressable by foreign commercial providers (e.g., a U.S. company will not be allowed to compete for the launch of an European Union precision navigation system and a Russian company will not be allowed to launch a U.S. military sat-

---

[23] Governments of the United Kingdom of Great Britain and Northern Ireland, the Union of Soviet Socialist Republics and the United States of America, "Treaty on Principles Governing the Activities of States in the Exploration and Use of Outer Space, Including the Moon and Other Celestial Bodies," New York, United Nations General Assembly, January 27, 1967. Liability for commercial satellites launched from an international spaceport is a recognized issue under space law and, in 2004, the United Nations adopted a resolution that recommends nation states "consider enacting and implementing national laws authorizing and providing for continuing supervision of the activities in outer space of nongovernmental entities under their jurisdiction;" include liability language in international launch agreements and "submit information on a voluntary basis on their current practices regarding on-orbit transfer of ownership of space object" (see United Nations, "Application of the Concept of the 'Launching State,'" New York, UN Resolution A/RES/59/115, January 25, 2005).

[24] Zelnio, 2006.

[25] Our definition of *addressable commercial launches* is different than that used by the FAA in their annual reports. The FAA definition includes all launches with a commercial payload or that are competed in the commercial market. Therefore, it includes many launches that are subject to international agreement such as travel from/to the ISS. Furthermore, the FAA definition does not recognize self-provisioning launches as nonaddressable.

ellite). If a nation does not have domestic launch capability, its national security launches may fall into another category, including addressable.

- **National affinity launches:** Although similar to the definition of national security launches, this category includes launches that are synonymous with national pride such as missions to the moon or interplanetary exploration. Again, if a nation does not have domestic launch capability, launch of these payloads may fall into another category, including addressable.

- **Self-provisioning:** This category includes launches where the satellite manufacturer or space service provider is also a launch provider (e.g., we assume SpaceX will launch its Starlink constellation on SpaceX launch vehicles). Increased vertical integration by aerospace companies is contributing to significant growth in this portion of the nonaddressable market.

Figure 1.4 plots the number of launches in the global heavy lift launch market in 1998, 2008, and 2018 as a function of the above categorizations. Although the global heavy lift launch market in total has grown significantly in the past decade, the portion over which commercial firms compete—the addressable share—has remained stagnant. Today, only 35 percent of the total market is addressable; nations, or launch providers themselves, increasingly fence off areas of the total market to support their own goals.[26]

It is this addressable share of the market that we have termed as the *global commercial market* in this report. The potential impact of national investments in domestic heavy lift launch capabilities on this commercial market is twofold: (1) The investments might shrink the addressable market, as emerging spacefaring nations begin to launch domestic payloads on domestic launch vehicles; and (2) more competitors might enter the commercial market, shrinking the share available to U.S. firms, as other nations also seek to offset their investments in heavy lift launch capabilities by selling to the global commercial market.

---

[26] For perspective, the average yearly nonaddressable market fenced off to support U.S. firms is 12 launches, or about 17 percent of the total market; the majority of which are in the national security category. We will examine the makeup of this share later when we examine the different strategies U.S. firms might use to position themselves in the global heavy lift launch markets.

Figure 1.4
Annual Global Heavy Lift Launches (1998, 2008, and 2018)

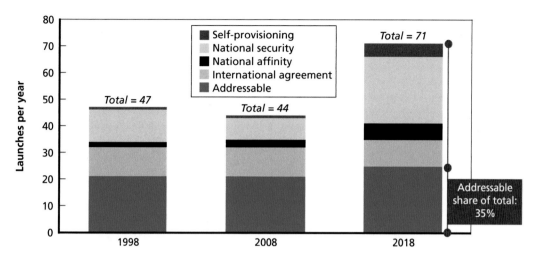

SOURCES: RAND analysis of FAA, *Commercial Space Transportation: 1998 Year in Review*, Washington, D.C., January 1999, and FAA *Commercial Space Transportation: 2008 Year in Review*, Washington, D.C., January 2009. The data for 2018 were derived from RussianSpaceWeb.com.

CHAPTER TWO

Understanding U.S. National Security Space Launch

In this chapter, we define the NSS heavy lift launch market. Globally, only a handful of launch vehicles can launch the mass-to-orbit required for U.S. NSS payloads.[1] For acquisition purposes, the USAF defines the NSS market in terms of mass-to-orbit requirements shown in Table 2.1, with the Polar 2 and Geosynchronous Direct Inject 2 being the most stressing.[2]

Space service providers use different orbits for different purposes. Figure 2.1 shows the primary orbits in use today. Low Earth sun-synchronous polar orbits are commonly used for Earth imaging satellites; medium Earth orbit (MEO) is used primarily for precision timing and navigation (e.g., GPS); and GEO has been the historical home for communication satellites serving regions below the 70th parallel. Communication satellites providing coverage to higher latitudes nearer the poles generally use a Molniya or Tundra highly elliptical orbit (HEO). GEO is matched to Earth's rotation rate such that satellites in that orbit appear stationary above a geographic region.[3] In all other orbits, satellites viewed from Earth cross our range of sight and then sunset beyond the Earth's limb.

Current Launch Vehicles Competitive in the Heavy Lift Launch Market

Comparing the requirements of Table 2.1 to the capabilities of global launch vehicles in the FAA's *Annual Compendium of Commercial Space Transportation: 2018*, we identified the launch vehicles in Table 2.2 as being competitive in one or more of the NSS

---

[1]   Not all NSS launches require heavy lift launch vehicles—but the majority do—and it is this class of launch vehicles for which the USAF asked RAND researchers to independently assess the market.

[2]   It is not simply the mass of the payload that makes a launch stressing, but also its destination. Elliptical transfer orbits are easier to achieve than circular orbits. The latter are termed *direct inject* because they are the final orbit of the payload. Payloads placed into a transfer orbit must carry their own fuel and propulsion systems, which are fired at apogee to circularize the orbit.

[3]   If the geosynchronous satellite is in an inclined orbit, it is not strictly stationary but appears to execute a small figure "8" motion above the geographic region.

Table 2.1
USAF Defined Mass-to-Orbit Requirements for National Security Space Acquisition

Reference Orbit	Apogee (nmi)	Perigee (nmi)	Inclination (deg)	Mass-to-Orbit (lbm)
LEO	500	500	63.4	15,000
Polar 1	450	450	98.2	15,500
Polar 2	450	450	98.2	37,500
MEO Direct Inject	9,815	9.815	50	11,750
MEO Direct Inject 2	10,988	10,988	55	9,000
MEO Transfer Orbit	10,998	540	55	9,000
MEO Transfer Orbit 2	10,988	540	55	11,200
Geosynchronous Transfer Orbit (GTO)	19,232	100	27	18,000
Molniya	21,150	650	63.4	11,500
Geosynchronous Direct Inject 1	19,323	19,323	0	5,000
Geosynchronous Direct Inject 1.5	19,493	19,493	0	8,000
Geosynchronous Direct Inject 2	19,323	19,323	0	14,500

NOTES: Apogee is the point in the orbit furthest from the Earth, perigee is the closest. Inclination is the angle from the equatorial plane. deg = degrees. lbm = pound-mass. nmi = nautical miles.

referenced orbits and as having been flown at least once in each of the past two years.[4] Four were manufactured in the United States, and two were made in Russia. China's Long March 3 A/B/E family of vehicles carry this class of payloads and are the workhorses of the Chinese launch market. The European Union, Japan, and India each have one vehicle competitive in this market, although the Indian vehicle is a relatively new entrant capable only of the LEO and Polar 1 mass-to-orbit requirements.

Near-Term Need for U.S. NSS-Certified Providers of Heavy Lift Launch

Although the United States currently has four launch vehicles competitive in one or more NSS-required orbital regimes, the Delta IV and the Atlas V (both marketed by

[4] FAA, 2018. The criteria for having at least one launch in each of the past two years excluded the Russian Angora and the Chinese Long March 2F, Long March 5, and Long March 7 from our study. Given the closed nature of their space and launch markets, it is difficult to assess if these launch vehicles are viable in the marketplace.

Figure 2.1
Characteristics of Common Satellite Orbit

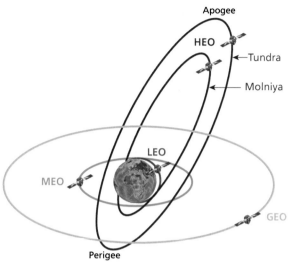

Orbit		Altitude (km)	
		Perigee	Apogee
LEO: low Earth orbit		200–2,000; normally: 600–1,000	
MEO: medium Earth orbit		2,000–GEO; normally: 10,000–20,000	
GEO: geostationary Earth orbit		35.786	
HEO: high elliptical orbit	Molniya (12 hours)	~500	~40,000
	Tundra (24 hours)	~24,000	~48,000

SOURCE: Simon Plass, Federico Clazzer, and Fritz Bekkadal, *Current Situation and Future Innovations in Arctic Communications*, in 2015 IEEE 82nd Vehicular Technology Conference (VTC2015-Fall), Boston: Institute of Electrical and Electronic Engineers, 2015.
NOTE: In their report, Plass, Clazzer, and Bekkadal define *GEO* as "geostationary Earth orbit" rather than "geosynchronous Earth orbit," which is the term we use in this report.

ULA) are not expected to remain in the market for NSS launch. The Delta IV is being retired from both commercial and NSS launch service, having insufficient orbital range to be economical.[5] The Atlas V has been the go-to launch vehicle for NSS, averaging four to six launches per year.[6] However, much of its capability is due to a first-stage engine of Russian manufacture (RD-180).[7] Congress has prohibited the USAF from purchasing additional Atlas V vehicles using the Russian engine after 2022.[8] Although

---

[5]   The Delta IV Heavy will be flown on NSS missions through 2024.

[6]   RAND analysis of historical NSS launch data.

[7]   *Aviation Week and Space Technology*, "First RD-180 Delivered," March 24, 1997.

[8]   National Defense Authorization Act for Fiscal Year 2017 §1602 (c)(2) reads:

   Contracts that are awarded during the period beginning on the date of the enactment of the National Defense Authorization Act for Fiscal Year 2017 and ending December 31, 2022, for the procurement of property or services for space launch activities that include the use of a total of 18 rocket engines designed or manufactured

Table 2.2
Global Launch Vehicles Currently Competitive in the Heavy Lift Launch Market

Country of Manufacture	Vehicle	NSS Required Orbital Regimes[a]	Total Launches in 2017 and 2018	Addressable Launches in 2017 and 2018
United States	Antares	LEO	3	0
United States	Delta IV	LEO, Polar 1 and 2, GTO	3	0
United States	Atlas V	LEO, Polar 1, GTO	11	1
United States	Falcon 9	LEO, GTO	38	26
Russia	Soyuz FG	LEO	9	0
Russia	Proton M	LEO	6	3
China	Long March 3 A/B/E	LEO	18	2
European Union	Ariane 5	LEO, Polar 1, GTO	12	10
Japan	H-II A/B	LEO	10	1
India	LVM3/GSLV-III	LEO, Polar 1	2	1

SOURCE: All data, except 2018 launches, are from RAND analysis of data in *The Annual Compendium of Commercial Space Transportation* (FAA, 2018). Since the 2019 compendium has not yet been published, 2018 launch data were verified against two online sources: RussiaSpaceWeb.com and SpaceFlightNow.com.

[a] Many of the launch vehicles listed in this table may be competitive in additional orbital regimes, but we were unable to substantiate that fact from the data contained in the FAA's 2018 compendium and were unwilling to take supplier marketing information as a valid source of data.

the Atlas V may continue to compete in the commercial market, its future is unclear and the number of engines on hand is limited.

To help renew competition in the NSS market, the USAF in 2011 created a certification path for new entrants to the NSS market, and the Falcon 9 was certified as an NSS supplier in 2015.[9] The Falcon 9, despite recent improvements in mass-to-orbit capabilities, falls short of the most stressing NSS requirements (i.e., Polar 2 and GEO Direct Inject 2). The final U.S. launch vehicle capable of servicing the NSS's LEO requirements, the Antares, uses Russian engines and has not been certified for NSS launch, competing instead for launches resupplying the ISS. The United States is currently in the process of certifying four new launch vehicles for NSS launch: SpaceX's Falcon Heavy, ULA's Vulcan, Northrop Grumman's OmegA, and Blue Origin's New Glenn.

---

in the Russian Federation, in addition to the Russian-designed or Russian-manufactured engines to which paragraph (1) applies (Public Law 114-328, 2016).

[9] Tracy Bunko, "New Entrant Certification Strategy Announced," Washington, D.C., Secretary of the Air Force Public Affairs, U.S. Air Force, October 14, 2011.

Future Launch Vehicles Planned to Be Competitive in the Heavy Lift Launch Market

Comparing the requirements of Table 2.1. with the capabilities of global launch vehicles in the FAA's *Annual Compendium of Commercial Space Transportation: 2018*, we identified the future launch vehicles in Table 2.3 as being competitive in one of more of the NSS referenced orbits.[10] Five are planned to be manufactured in the United States, but only one has had a first launch. China is planning for two entrants, both of which had a first launch in 2016 but have been largely dormant since. Russia, the European Union, and Japan each have one entrant, none of which have currently launched. Hypothetically, if all of these launch vehicles were to mature as planned, in 2022 there would be an astounding 19 vehicles competitive in the heavy lift launch market.[11] However, we caution that expected first launch dates listed here are the earliest launch dates and that it often takes several years from first launch before a launch vehicle gains a significant share of the global market, even with government sponsorship or assistance.

Table 2.3
Future Launch Vehicles Competitive in the Heavy Lift Launch Market

Country of Manufacture	Vehicle	NSS Required Orbital Regimes	Actual or Expected First Launch
United States	Falcon Heavy	All	2018
United States	Falcon Super Heavy	All	2020
United States	New Glenn	All	2021
United States	OmegA	GTO	2021
United States	OmegA Heavy	All	2024
United States	Vulcan	All	2021
Russia	Irtysh (Soyuz 5)	LEO (likely all)	2022
China	Long March 5	LEO, GTO	2016
China	Long March 7	LEO, Polar 1	2016
EU	Ariane 6	LEO, GTO	2020
Japan	H-III	LEO	2020

SOURCE: RAND analysis of data in *Annual Compendium of Commercial Space Transportation: 2018* (FAA, 2018). Planned first launch dates for vehicles is taken from USAF NSS Certification plans or, for non-NSS–certified vehicles, the most recent manufacturer press releases as of June 2019.

---

[10] FAA, 2018.

[11] This count assumes the Delta is retired, but all other current launch vehicles remain in the market.

Status of Efforts to Certify Providers for NSS Launch

In 2014, the USAF developed an incremental approach to achieve competition in NSS launch. The Falcon 9 was then being certified for NSS launch and the plan leveraged that fact to propose a three-phased approach as described in Figure 2.2 by the U.S. Government Accountability Office (GAO) in its 2015 analysis of the plan. Phase 1A, enabled by the expected NSS certification of the Falcon 9, would compete up to 14 launches between SpaceX's Falcon 9 and ULA's Delta IV and Atlas V vehicles over the next three years. Phase 2 would begin in 2018 using an unspecified acquisition approach, followed by Phase 3 beginning in 2023.[12]

The USAF released a launch service agreement (LSA) RFP on October 5, 2017, to transition from the use of Russian engines and to implement affordable assured access to space via sustainable competition with commercial launch providers.[13] Ultimately, three contracts were awarded in October 2018 to mature the designs of the Vulcan, New Glenn, and OmegA launch vehicles.[14]

Figure 2.2
GAO's 2015 Depiction of USAF Approach to Achieving Competition in the NSS Launch Market

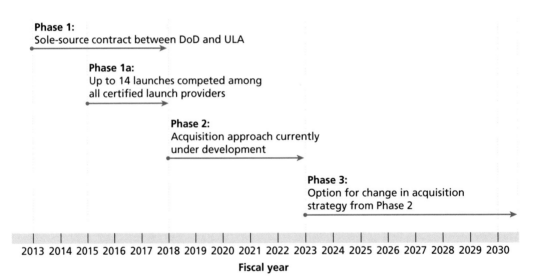

SOURCE: GAO, *Evolved Expendable Launch Vehicle: The Air Force Needs to Adopt an Incremental Approach to Future Acquisition Planning to Enable Incorporation of Lessons Learned*, Washington, D.C., GAO 15-623 August 2015.

---

[12]   Phase 2 is currently being completed and expected first launch dates are set for fiscal year 2022.

[13]   Department of the Air Force, Air Force Space Command, "FA8811-17-9-0001; Evolved Expendable Launch Vehicle (EELV) Launch Service Agreements (LSA) Request for Proposals (RFP)," Solicitation Number FA8811-16-R-000X, March 2017.

[14]   In 2019, Elon Musk revealed that SpaceX had submitted a proposal for maturation of the Falcon Heavy but had not received an award.

In concert with those LSA efforts, the USAF in 2016 funded Rocket Propulsion System prototypes to mature engine development for future launch vehicles. Awards were made to SpaceX for development and testing of the Raptor engine (used in the Super Heavy); to Orbital ATK (now Northrop Grumman Innovation Systems [NGIS]) for development of the Graphite Epoxy Motor (GEM) 63XL strap-on booster (to be used in both the OmegA and Vulcan), as well as Castor 300 and 600 engines; and Aerojet Rocketdyne for development of the AR1 booster. Blue Origin subcontracted under awards for Orbital ATK to mature an extendable nozzle for the BE-3U/EN upper-stage engine (to be used on the New Glenn) and for ULA to develop the BE-4 first-stage engine (to be used on the New Glenn and Vulcan). SpaceX had also received funding from NASA for the original development of the Falcon 9, and both SpaceX and NGIS have launch contracts with NASA in support of the ISS.[15] Total U.S. government funding of U.S. heavy lift launch providers since 2005 is shown in Figure 2.3.

For the four U.S. launch vehicles currently seeking NSS certification, Figure 2.4 shows current and planned milestones. To allow the broadest participation in the NSS market, the NSS Certification Guide provides multiple paths to certification from which a supplier can select when submitting its letter of intent for certification to the USAF.[16] Three of the vehicles (New Glenn, OmegA, and Vulcan) have chosen the path where the USAF participates in the complete cycle of preliminary and critical design reviews (PDRs and CDRs) throughout each vehicle's development and is then certified after two successful launches. The fourth vehicle (Falcon Heavy) has chosen a path whereby the USAF participates in a series of engineering review boards (ERBs) to assess the vehicle's risks related to critical components, and then the vehicle will be certified after completing three successful launches.

In Figure 2.4, we also plotted the projected first launch dates of the Falcon Heavy, which displays a fairly typical pattern for first launch forecast as being "one year away" over a number of years leading up to a successful first launch in the first quarter of 2018. As one experienced industry veteran observed, the same pattern would be seen if we were to go back to the late 1970s and plot the forecasts for the first launch of the space shuttle. We will say more about the accuracy of first launch predictions in our chapter on assessing the impact of USAF acquisition decisions on the NSS market.

---

[15] NGIS supports the ISS using its Antares launch vehicle.

[16] USAF, *Launch Services New Entrant Certification Guide*, Washington, D.C.: U.S. Accountability Office, GAO 13-317R, February 2013.

Figure 2.3
Total U.S. Government Funding of U.S. Providers of Heavy Lift Launch Vehicles Since 2005 (in 2018 Dollars, Billions)

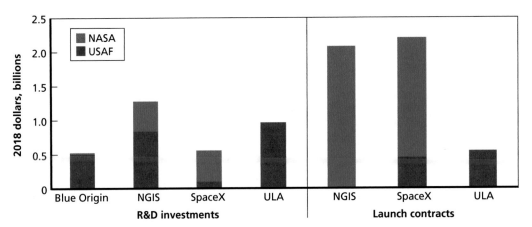

SOURCE: RAND analysis of Irene Klotz and Jen DiMascio, "SpaceX Loses Out on U.S. Air Force Next-Gen Launcher Development," *Aviation Week and Space Technology*, October 12, 2018; Jeff Foust, "Air Force Adds More Than $40 Million to SpaceX Engine Contract," *SpaceNews*, October 21, 2017; NASA, *Blue Origin Space Agreement*, Washington, D.C.: NASA Crew and Cargo Program Office, 2010; *Aviation Week and Space Technology*, "SpaceX Wins NASA Launch Services Contract," April 24, 2008; NASA, "Commercial Orbital Transportation Services a New Era in Spaceflight," 2014; NASA, *National Aeronautics and Space Administration FY 2018 Agency Financial Report*, Washington, D.C., 2018; Space and Missile Systems Center, "Air Force Awards Two Rocket Propulsion System Prototype OTAs," 2016b; Space and Missile Systems Center, "Air Force Awards Final Rocket Propulsion System Prototype OTAs," 2016a.

Figure 2.4
NSS Certification Timelines of New Entrants to the Heavy Lift Launch Market

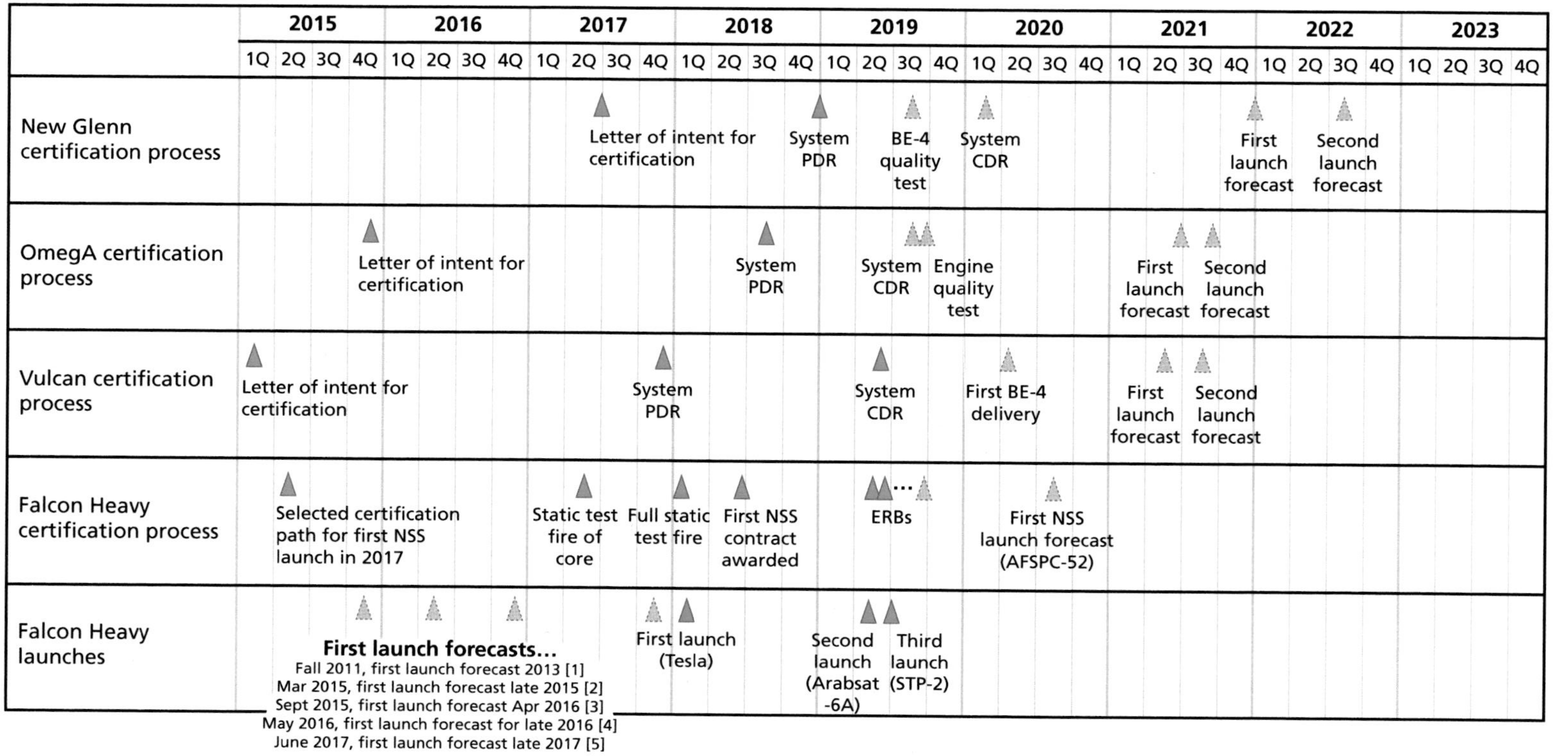

Process	2015	2016	2017	2018	2019	2020	2021	2022
New Glenn certification process			Letter of intent for certification	System PDR	BE-4 quality test	System CDR	First launch forecast	Second launch forecast
OmegA certification process		Letter of intent for certification		System PDR	System CDR	Engine quality test	First launch forecast / Second launch forecast	
Vulcan certification process	Letter of intent for certification		System PDR		System CDR	First BE-4 delivery	First launch forecast / Second launch forecast	
Falcon Heavy certification process	Selected certification path for first NSS launch in 2017		Static test fire of core / Full static test fire	First NSS contract awarded	ERBs	First NSS launch forecast (AFSPC-52)		
Falcon Heavy launches	First launch forecasts...		First launch (Tesla)	Second launch (Arabsat-6A)	Third launch (STP-2)			

First launch forecasts...
Fall 2011, first launch forecast 2013 [1]
Mar 2015, first launch forecast late 2015 [2]
Sept 2015, first launch forecast late 2016 [3]
May 2016, first launch forecast for late 2016 [4]
June 2017, first launch forecast late 2017 [5]

SOURCES: RAND analysis of information from SpaceX, "SpaceX Announces Launch Date for the World's Most Powerful Rocket," press release, April 5, 2011; Peter B. de Selding, "SpaceX Aims to Debut New Version of Falcon 9 This Summer," *SpaceNews*, March 20, 2015; Jeff Foust, "First Falcon Heavy Launch Scheduled for Spring," *SpaceNews*, September 2, 2015; Mike Wall, "More Power! SpaceX's Rockets Are Stronger Than Predicted," Space.com, May 2, 2016; Caleb Henry, "SpaceX's Final Falcon 9 Design Coming This Year. Two Falcon Heavy Launches Next Year." *SpaceNews*, June 27, 2017.

CHAPTER THREE

Understanding the Addressable Commercial Launch Market

In this chapter, we take a historical look at the size and dynamics of the addressable commercial market for heavy lift launch capabilities, the sources of supply, the drivers of demand, factors that affect buyers' selection of launch vehicle providers, and how selection as a NSS launch provider may change the competitiveness of U.S. firms in the commercial market.

Dynamics of Supply in the Addressable Commercial Launch Market

The size of the addressable commercial heavy lift launch market has been relatively stable over the past 12 years with an average of approximately 20 launches a year (see Figure 3.1).

To understand the historical dynamics and potential future of the market, we first characterize the sources of supply for heavy lift launch services and then the drivers of demand for those services.

Change in a Decade of Stable Market Shares as SpaceX Enters and Sea Launch and Proton Struggle with Launch Failures

Segmenting the addressable commercial market by launch vehicle reveals how heavy lift launches have been supplied over the last decade. In the early to mid-2000s, the market was split fairly evenly between Sea Launch, Proton, and Arianespace. However, in 2007, Sea Launch experienced its first launch failure and was forced to file for bankruptcy in 2009. Sea Launch continued to struggle to compete in the market in the early 2010s and was ultimately forced out of the market after a second launch failure in 2013. Proton also experienced a string of launch failures from 2006 to 2016 with approximately one per year. In contrast to Sea Launch, Proton's share of the addressable commercial market was not significantly affected until 2016. It made no commercial launches in 2018. Proton is scheduled to launch a payload for Eutelsat in 2019 and may continue to compete in the commercial market despite its many launch failures.

As Sea Launch and Proton began to struggle with launch failures in the mid-2000s, SpaceX began to explore the potential of entering the commercial market with

Figure 3.1
Global Addressable Commercial Heavy Lift Launch Market by Launch Vehicle, 2007–2018

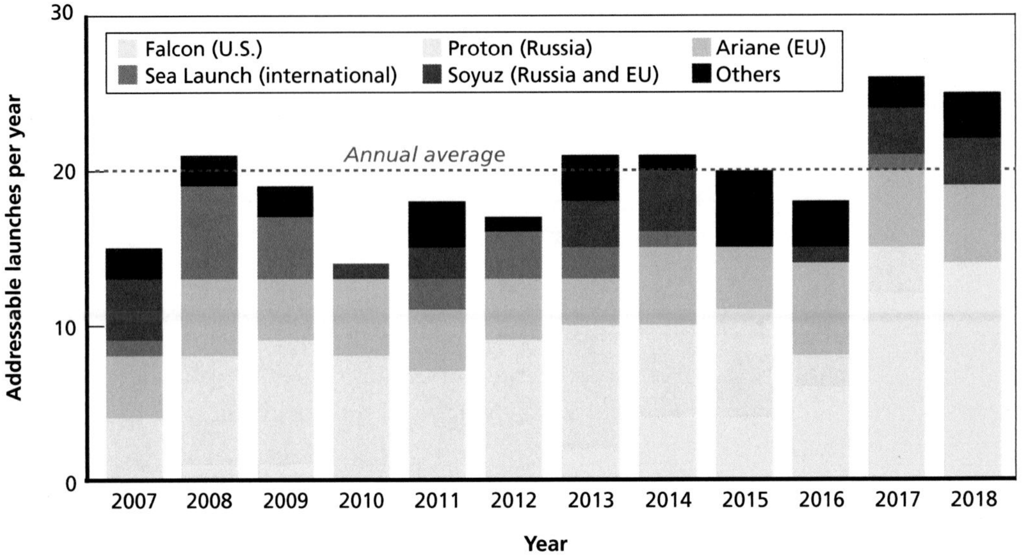

SOURCE: RAND analysis of heavy lift launch data set, 2007–2018 (Anatoly Zak, "Russian Space Program and Rocket Development in 2018," *RussianSpaceWeb.com*, August 6, 2019b; Union of Concerned Scientists, "UCS Satellite Database," webpage, last updated March 31, 2019; U.S. Department of Transportation, Bureau of Transportation Statistics, "Table 1-39: Worldwide Commercial Space Launches," webpage, last updated May 21, 2017; World Bank, "World Bank Country and Lending Gap," webpage, undated; World Bank, "Classifying Countries by Income," webpage, October 4, 2018).

the Falcon 1 and made one launch in both 2008 and 2009. In 2010, SpaceX made its first successful launch of the Falcon 9 and, in 2012, began providing NASA with Commercial Orbital Transportation Services (COTS) to resupply the ISS. Having developed confidence in the Falcon 9 by providing launch services to NASA and having built up significant launch capacity, SpaceX positioned itself well to move into the commercial market in 2013 as Sea Launch was forced out and Proton's string of launch failures continued to accumulate. Since 2013, SpaceX's share of the global commercial launch market has continued to grow and has reached more than 50 percent in 2018, having taken much of Sea Launch and Proton's former market shares.

Over the past decade, Arianespace has continued to be a stable competitor in the commercial heavy lift launch market with an average of four to five launches a year. Soyuz has also continued to have a presence in the commercial market with approximately two launches a year, but with less stability than Arianespace.

Overall, the reliability of launch vehicles has driven much of the supply dynamics in the heavy lift launch market over the past decade. The Falcon 9 and Arianespace continue to be highly reliable and have been the most-competitive launch vehicles in the commercial market in recent years. In contrast, a string of launch failures for

Proton and Soyuz diminished the previously dominant market share of the former and has led to unstable demand for the latter. However, Russia's commitment to maintaining national launch capabilities, and to these launch vehicles in particular, seems to have heavily influenced the ability of these vehicles to remain in the market. Sea Launch, on the other hand, did not have the same governmental support and was quickly forced to file for bankruptcy after its first launch failure and was propelled out of the market entirely after its second.

Assessment of Foreign Suppliers' Status in the Commercial Heavy Lift Launch Market

U.S. suppliers of heavy lift launch services are facing strong international competition in the global market and likely will continue to do so. Figure 3.2 illustrates the current global market for competitive heavy lift launch vehicles, which includes Arianespace's Ariane 5, Russia's Soyuz, China's Long March 3A/B/E, Japan's HII-A/B, and India's GSLV-III.

Arianespace has historically been a strong competitor in commercial heavy lift launch services but has lost some market share since the introduction of SpaceX's Falcon 9 in 2013. During this period, Russia's commercial launch suppliers have also experienced financial and organization turmoil. However, the Soyuz has been a consistent supplier of heavy lift launch services for more than 20 years and, because it is the only launch vehicle currently capable of transporting crew to the ISS, will likely con-

Figure 3.2
Positioning of Competitors in the Global Commercial Heavy Lift Launch Market

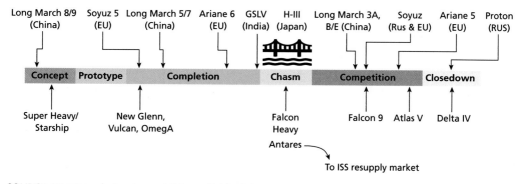

SOURCE: RAND analysis using publicly available data.
NOTE: The Long March 5/7 position on the timeline is due to the fact that Long March 5 had one partially successful launch in 2016 and one failure in 2017. The Long March 7 had successful launches in 2016 and 2017. However, neither vehicle has launched since then. Also, note that this figure is not our analysis of where U.S. competitors are positioned in the NSS-certified launch market. Given the ongoing USAF source selections, we have portrayed new U.S. entrants at the same level so as not to give the appearance of preference. The Falcon Heavy position at the Chasm level is based on completion of three successful launches as of June 2019, one of which was commercially competed.

tinue to compete in the global market. Moreover, Russia is investing in a new launch vehicle, the Irtysh/Soyuz 5, and is likely to regain its historical market share after the vehicle's expected introduction in 2022. Similarly, Arianespace is investing in a new launch vehicle, the Ariane 6, and has announced that OneWeb signed a contract to be its anchor customer in 2020. Thus, Arianespace is also positioning itself to regain its historical market share and further diminish that of U.S. firms, which is currently well above the historical average.

China, Japan, and India have not had a strong presence in the commercial heavy lift launch market. However, each has demonstrated some interest in competing further and should be monitored closely. Although largely shut out of the commercial market because of the U.S. ITAR and Arms Export Control Act, China has begun to provide launch services through its Belt and Road Initiative[1] and may try to compete for more launch contracts with less traditional buyers. Although Japan does not actively compete in the commercial market, it recently accepted two launch contracts from Inmarsat and may continue to accept such contracts given the current lack of providers in the market.[2] Both China and Japan are developing new launch vehicles (the Long March 5/7 and the H-III, respectively) and may be positioning themselves to compete more effectively in the heavy lift launch market over the long term. India had its first commercial heavy lift launch in 2018 and continues to demonstrate a strong desire to compete.

Dynamics of Demand in the Addressable Commercial Launch Market

Communication Service Providers Have Been the Dominant Users of Commercial Heavy Lift Launch Services

Figure 3.3 depicts the addressable market segmented by primary payload type. It clearly shows that communication service providers have historically been and continue to be the dominant buyers of commercial heavy lift launch services. In contrast, imaging and earth science satellite providers, the second and third largest users of heavy lift launch capabilities, have procured only six and four launches, respectively, over the past 12 years.

In recent years, the telecommunications industry has encountered several disruptions brought on by the increasing use of optical fiber, the advent of 5G technology, and new proposals for highly proliferated constellations of small satellites in LEO. Some traditional communications service providers with large constellations in GEO have begun to reevaluate their market strategies. Space service providers, such as

---

[1]   China launched a communications satellite for Algeria as part of its Belt and Road Initiative in 2017. China has begun bundling satellites and launch services along with financial incentives.

[2]   Japan does not currently have an organization for marketing its launch services to commercial buyers.

Figure 3.3
Global Addressable Commercial Heavy Lift Launch Market by Primary Payload Type, 2007–2018

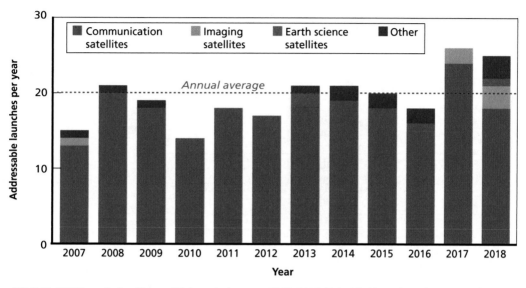

SOURCE: RAND analysis of heavy lift launch data set, 2007–2018 (Zak, 2019b; Union of Concerned Scientists, 2019; U.S. Department of Transportation, Bureau of Transportation Statistics, 2017; World Bank, undated; World Bank, 2018).

DirecTV, have announced their intention to exit the satellite communications market entirely. Others, such as SES, are diversifying their fleets with constellations of smaller satellites in other orbits. There also are a few service providers, such as Intelsat, Telesat, and Echostar, that are at points in their capital investment cycle when they do not need to make immediate decisions about their future strategies and are waiting to see how the market evolves before committing to a course of action. Finally, there are a relative few space-based communication service providers, such as Eutelsat and Inmarsat, who remain committed to the market and plan to launch new satellites in the coming years.

Past and future launches of the seven largest satellite communication constellations in GEO (i.e., Intelsat, SES, Eutelsat, Echostar, Telesat, JCSAT, and Inmarsat) are illustrated in Figure 3.4. Demand from these seven communications service providers is cyclical and represents approximately 50 percent of the total launches from the telecommunications industry in the last decade. However, only three of the seven service providers (i.e., Intelsat, Eutelsat, and Inmarsat) have announced plans to launch satellites after 2019. This could be indicative of disruptions in the industry. It is also possible that the small number of announced launches is not representative of the actual demand. Or it might be the case that the smaller demand is representative but reflects the longer lifetimes of newer satellites and thus longer time gaps (e.g., cyclical period) between launches for these service providers. For these reasons, we did not use the data

Figure 3.4
Launches for Large and Midsized GEO SATCOM Constellations, 2007–2021

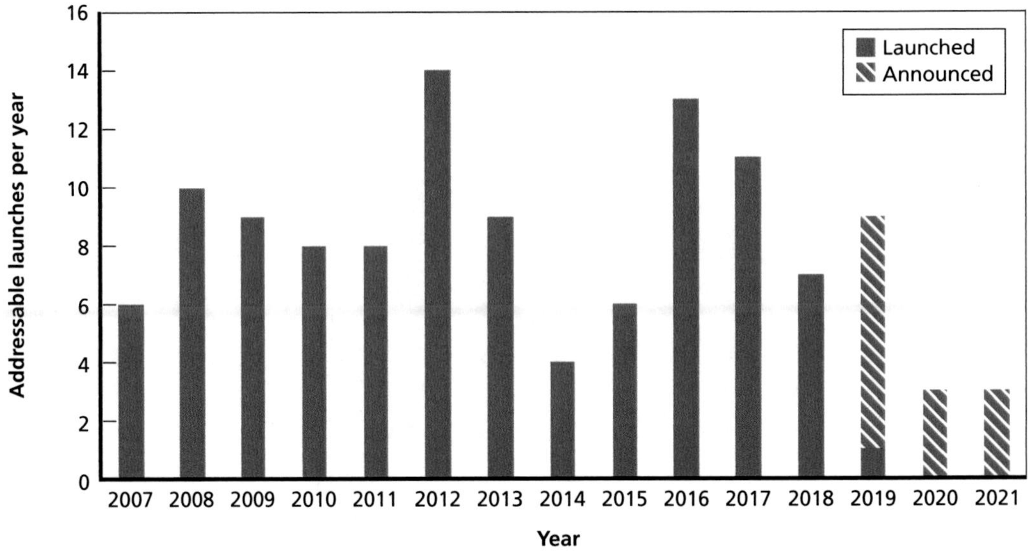

SOURCE: RAND analysis of heavy lift launch data set 2007–2018 (Zak, 2019b; Union of Concerned Scientists, 2019; U.S. Department of Transportation, Bureau of Transportation Statistics, 2017; World Bank, undated; World Bank, 2018).

in Figure 3.4 as the basis for our forecast of future demand in the addressable launch market.

GEO Has Been the Primary Destination for Heavy Lift Launches, but LEO May Be on the Rise

Figure 3.5 segments the addressable market by orbit. It reveals that GEO has been the primary destination of satellites launched on heavy lift launch vehicles. In 2017 and 2018, the number of heavy lift launches to LEO rose, but half of these launches were by SpaceX to build out Iridium's new satellite communications constellation that was completed in the first quarter of 2019. For heavy lift launches to MEO, Arianespace executed the handful of launches during this period to begin building out SES's new O3b constellation. These LEO and MEO launches do not represent new sources of demand, because they were procured by traditional buyers of heavy lift launch services. A new destination does not translate to a need for additional launch services, as we will discuss in the next section. However, if some of the new companies with proposed proliferated LEO and MEO constellations come to fruition and are initially built up, there may be a slightly higher demand for a few years than has been observed historically.

Figure 3.5
Global Addressable Commercial Heavy Lift Launch Market by Orbit, 2007–2018

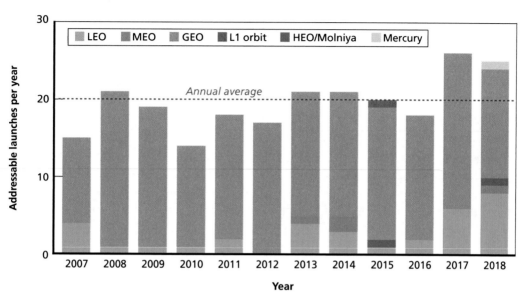

SOURCE: RAND analysis of heavy lift launch data set 2007–2018 (Zak, 2019b; Union of Concerned Scientists, 2019; U.S. Department of Transportation, Bureau of Transportation Statistics, 2017; World Bank, undated; World Bank, 2018).

Demand from Proliferated Constellations of Small Satellites Is Expected to be Small

It has been stated that the demand for launch services will grow significantly in the coming years as such companies as SpaceX, OneWeb, Telesat, and Amazon begin to execute their plans for highly proliferated constellations of small satellites in LEO and MEO. However, it is important to assess how many and what types of launch services are likely to be procured and who the launch service providers are likely to be. Heavy lift launch is the preferred method for building out these initial constellations because it provides an efficient and cost-effective means of populating entire planes within a constellation, requiring fewer launches. However, small and medium lift launches may become the preferred methods for replenishing and/or replacing satellites once initial constellations are completed because of variable demand in the number and orbits of the satellites.

To better understand how these constellations might affect the heavy lift launch market, we examined the Iridium Next constellation build-out. In the first quarter of 2019, Iridium completed its new constellation of 75 satellites. It procured launch services from SpaceX and launched approximately one vehicle a quarter, each with ten satellites, from the beginning of 2017 to the beginning of 2019, as illustrated in

Figure 3.6.[3] This amounted to a moderate demand of 7.5 launches over two years. However, Iridium is unlikely to generate significant demand for heavy lift launch over the next decade, because the lifetime of the new constellation is expected to be approximately 15 years.

Taking a closer look at the plans for SpaceX, OneWeb, Telesat, and Amazon, Table 3.1 summarizes the number of satellites, approximate number of launches, and the announced launch providers for these constellations. Launches for two of these constellations are unlikely to be competed in the commercial market. SpaceX is a vertically integrated space service and launch service provider and plans to self-provision launches for their Starlink constellation. Similarly, Amazon's Project Kuiper will likely launch its satellites with services provided by its sister company Blue Origin and thus, effectively, be nonaddressable. Launch service providers for the OneWeb and Telesat constellations have also been announced. For OneWeb, Arianespace and Blue Origin will be the primary launch providers, and each will likely provide four to six launches each over two to three years. For Telesat, Blue Origin will be the primary launch pro-

Figure 3.6
Iridium Next Constellation Build Out, First Quarter 2017 to First Quarter 2019

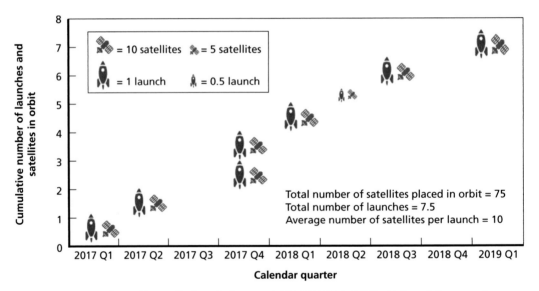

SOURCE: RAND analysis of heavy lift launch data set, 2007–2018 (Zak, 2019b; Union of Concerned Scientists, 2019; U.S. Department of Transportation, Bureau of Transportation Statistics, 2017; World Bank, undated; World Bank, 2018).

---

[3]   The sixth launch in the second quarter of 2018 had only five Iridium satellites, but it was partnered with various other small satellites.

Table 3.1
Plans for Proliferated Satellite Constellations

Company	IRL	Orbit	Number of Satellites	Satellites per Launch	Launches	Launch Provider	Planned Initial Service Date
OneWeb[a]	Chasm	LEO	720	~30	24	Arianespace and Blue Origin	2021
		MEO	1,280		42		Unknown
Telesat[b]	Prototype	LEO (Ka-band)	117	~10	12	Blue Origin	2022
		LEO (V-band)	117		12		Unknown
SpaceX[c]	Prototype	LEO	4,409	~60	74	Self-provisioned	2020–2021
		VLEO	7,518		126		Unknown
Amazon	Concept	LEO	3,236	Unknown	Unknown	Self-provisioned[d]	Unknown

SOURCES: Amazon's plans for a proliferated LEO constellation as submitted to the Federal Communications Commission (FCC) for approval in July 2019; FCC approval for the second phase of OneWeb's constellation for satellites in MEO is pending.

[a] OneWeb launched the first six satellites of its LEO constellation using a Soyuz rocket provisioned by Arianespace on February 27, 2019. OneWeb also plans to use Arianespace's Ariane 6 and Blue Origin's New Glenn to build out its initial constellation. For replenishment and replacement, OneWeb plans to use Virgin Orbit's LauncherOne, a small air-launched vehicle.

[b] Telesat has launched two test satellites (the first on November 28, 2017) on a heavy lift Russian Soyuz (colaunched with nine other small satellites), but it was lost because of an issue with the rocket's upper stage, and the second on January 12, 2018, on a medium lift Indian PSLV (colaunched with 31 other small satellites).

[c] SpaceX launched the first 60 satellites of its Starlink constellation on May 24, 2019, and has plans for two more launches in 2019.

[d] Amazon's Project Kuiper is likely to be "self-provisioned" by Blue Origin, which is also owned by Jeff Bezos.

vider and likely will provide four to six launches over two to three years. Thus, only the OneWeb constellation is generating new demand for addressable heavy lift launch services. Telesat has simply shifted its historical demand to a new orbit and the other two constellations are self-provisioned.

Moreover, these proliferated constellations are at very low IRLs, with only OneWeb testing the business model at the time of the writing of this report.[4] The business cases for these systems are reliant on cheap satellites and cheaper launches, but they are also reliant on demand for satellite internet. SpaceX's Starlink system has demonstrated a failure rate that threatens space sustainability; improving the failure

[4] In early 2019, OneWeb changed its business model to adapt to the realities of the space-enabled services market (see Andy Pasztor, "Startup Satellite Venture OneWeb Blasts Off With Revised Business Plan," *Wall Street Journal*, February 27, 2019).

rate to an acceptable level is likely to increase costs.[5] Demand for satellite internet has so far failed to meet the expectations of current providers. The difference between the number of people without internet and the number of people willing to pay for internet may be much larger than these business models anticipate.[6]

Demand for Heavy Lift Launch Services from High-Income Nations Is Cyclical, and It May Be Slowly Growing from Middle- and Lower-Income Nations

Segmenting the addressable market by national income of space service operators, we observe distinct dynamics among high-income nations versus middle- and lower-income nations.[7]

The demand for heavy lift launch services from space service operators in high-income nations, plotted in Figure 3.7 is cyclical but stable. The stability in demand, with an average of 15 launches a year, demonstrates the underlying market for satellite communications is mature. The cyclical nature of the demand is driven by the capital investment cycle for these space service operators, which involves replacing or replenishing satellites approximately every five years.

In contrast, the demand from space service operators in middle and lower income nations is slowly growing, with an average of approximately three launches a year from 2007 to 2012 and approximately six launches a year from 2013 to 2018. Slow growth in this segment of the market also demonstrates that the satellite communication market is maturing as space service operators from middle- and lower-income nations are entering the market as late adopters with fewer resources and perhaps greater risk aversion than higher-income nations. Russia and China make up 40 percent of this growing demand, while Brazil, India, Indonesia, and Malaysia each were the beneficiaries of at least three heavy lift launches in the past 12 years. All middle- and lower-income nations that are operating payloads that used heavy lift launch services over the past decade are summarized in Table 3.2.

U.S. launch service providers have not traditionally pursued this segment of the heavy lift launch market and may need to look outside their normal customer base if they are to benefit from this growth. Russia, China, India, or Japan may be better positioned to serve this segment of the market.

---

[5]   Jonathan O'Callaghan, "'Not Good Enough'—SpaceX Reveals That 5% Of Its Starlink Satellites Have Failed In Orbit So Far," *Forbes*, June 30, 2019. For perspective, a 5-percent failure rate of a 12,000 satellite constellation creates 600 new pieces of debris; this is equivalent to ten times the long-term debris caused by the Indian ASAT test as of July 2019.

[6]   Daniel Oberhaus, "SpaceX Is Banking on Satellite Internet. Maybe It Shouldn't," *Wired*, May 19, 2019.

[7]   National income designations are developed by the World Bank and updated annually. For more information and an interactive tool, see World Bank, "Classifying Countries by Income," webpage, October 4, 2018.

Figure 3.7
Annual Number of Addressable Heavy Lift Launches Segmented by Income of the Operating Country: World Bank–Designated Income Categories, 2007–2018

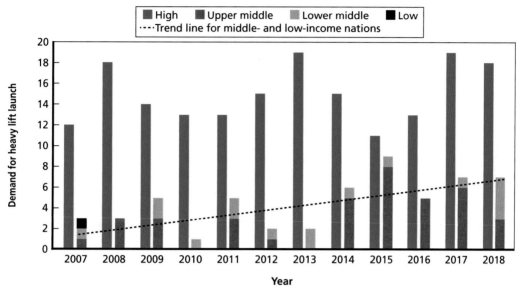

SOURCE: RAND analysis of heavy lift launch data set 2007–2018 (Zak, 2019b; Union of Concerned Scientists, 2019; U.S. Department of Transportation, Bureau of Transportation Statistics, 2017; World Bank, undated; World Bank, 2018).

RAND's Forecast of Global Demand for Heavy Lift Launch Services Addressable by U.S. Firms

We forecasted the global demand for heavy lift launch services in the commercial market based on the dynamics observed over the past decade. The first component of our forecast was the stable but cyclical demand from high-income nations driven primarily by capital investment cycles in satellite communications, and the second was slow growth in demand from middle- and lower-income nations as their use of heavy lift launch services increase. Our forecast was generated using a Monte Carlo simulation that assumes a normal distribution of annual launches based on the cyclical behavior[8] and linear

---

8 Cyclical demand is modeled as

$$A_{HIN} \cos\left[\frac{2\pi}{\tau_{HIN}}\left(t - t_0\right)\right].$$

Table 3.2
World Bank-Designated Middle- and Lower-Income Nations Operating Payloads Launched by Heavy Lift Launch Vehicles, 2007–2018

Year	Upper Middle	Lower Middle	Lower	Year	Upper Middle	Lower Middle
2007	Russia	China	Nigeria	2013		India Bolivia
2008	Venezuela Russia Brazil			2014	Turkey Thailand	Malaysia China	Egypt
2009	Russia Malaysia	Indonesia China		2015	Turkmenistan Turkey Russia	Mexico China	Laos
2010		Egypt		2016	Thailand Russia China	Brazil Belarus	
2011	Russia China	Pakistan Nigeria		2017	Russia China Bulgaria	Brazil Algeria	Angola
2012	China	Indonesia		2018	Russia China		Indonesia India Bangladesh

SOURCES: RAND analysis of heavy lift launch data set, 2007–2018 (Zak, 2019b; Union of Concerned Scientists, 2019; U.S. Department of Transportation, Bureau of Transportation Statistics, 2017; World Bank, undated; World Bank, 2018).
NOTE: The World Bank's national income designations are updated yearly. Therefore, China transitions from lower-middle to upper-middle income between 2009 and 2011. Similarly, Nigeria transitions from lower to lower-middle income between 2007 and 2011.

growth[9] in historical demand extrapolated into the future. The uncertainty bounds of the forecast are calculated based on the standard deviation of the historical demand, which increases as the forecast progresses in time.

Our forecast, illustrated in Figure 3.8, may be pessimistic if global demand for heavy lift launch from lower and middle income nations accelerates. However, the long-term trajectory of the telecommunications industry has yet to be decided. If the industry simply moves away from traditional constellations of fewer large satellites in GEO to constellations of more small satellites in LEO and MEO, demand could potentially remain stable because of the launch requirements for building out these constellations. If instead these firms exit the space-enabled services market altogether,

---

[9] Linear growth is modeled as

$$m_{MLIN}\left(t-t_0\right)-b_{MLIN}.$$

Figure 3.8
Global Addressable Heavy Lift Launch Market, Historical, 2007–2018, and RAND Forecast, 2019–2030

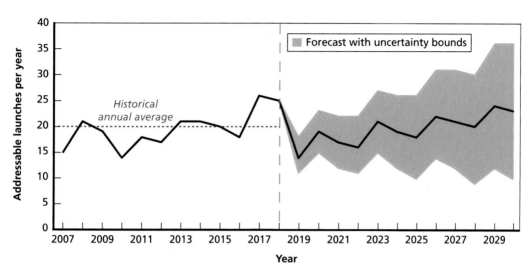

SOURCE: RAND analysis of heavy lift launch data set, 2007–2018 (Zak, 2019b; Union of Concerned Scientists, 2019; U.S. Department of Transportation, Bureau of Transportation Statistics, 2017; World Bank, undated; World Bank, 2018).
NOTE: The cyclical component is estimated based on the historical addressable commercial demand from high income nations, and growth is estimated based on the addressable commercial demand from middle- and lower-income nations.

our forecast may be optimistic.[10] On balance, we believe our forecast to be neither optimistic nor pessimistic.

Compared with the historical range of the addressable market and with forecasts we obtained from other sources, our forecast appears reasonable. The forecasts we compared against are the FAA's annual forecast (the latest of which are those published in summer 2017 and summer 2018), a launch service provider, and an insurer. The latter two forecasts were provided to us on a not-for-attribution basis and are therefore referred to as Insurer A and Supplier A. As shown in Figure 3.9, our forecast falls below the near-term forecasts (i.e., 2019 to 2021) from the FAA and Insurer A—however, both the insurer and the FAA acknowledge that their forecasts are traditionally higher than what is achieved, so this is as we would expect.[11] The near-term forecast

---

[10] Disruption in the telecommunications industry, historically the dominant procurer of commercial heavy lift launch services, is causing firms to reevaluate their market strategies. Some have decided to no longer invest in space-enabled services, while others continue to invest. Some buyers we interviewed noted that their capital expenditure cycle allows them to wait before making decisions, so near-term declines may not be indicative of future demand.

[11] An assessment of historical FAA forecasts revealed only 77 (±16 percent) of forecasted launches actually materialized. Insurer A and FAA forecasts are for a larger market than the addressable launches; there may be small error in our adjustment of them to proportionally reflect the smaller size of the addressable market. Given that

Figure 3.9
Comparison of Global Addressable Heavy Lift Launch Market Forecasts, 2017–2030

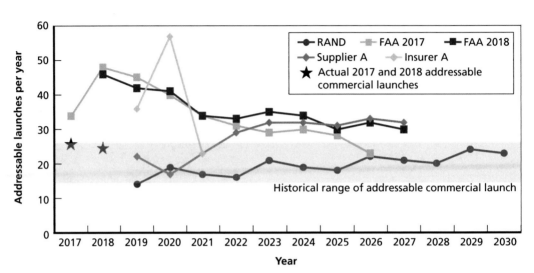

SOURCES: RAND forecast = RAND analysis. FAA 2017 forecast = FAA, *The Annual Compendium of Commercial Space Transportation: 2017*, Washington, D.C., January 2017. FAA 2018 forecast = FAA, 2018. Supplier A forecast = provided to RAND on a nonattribution basis. Insurer A forecast = provided to RAND on a nonattribution basis. Historical range = RAND analysis.

from Supplier A is comparable and uses the same definition of the addressable market that is used in this report. Supplier A uses an adjusted FAA forecast for years 2022 and beyond.

Factors Commercial Buyers Consider when Selecting a Launch Service Provider

The communications service market, which is the predominant commercial buyer of heavy lift launch, is characterized by very large upfront capital investments, followed by long revenue streams over the lifetime of the satellites. In our interviews, commercial buyers stated they select launch service providers based on reliability and insurability; scheduling flexibility; and launch vehicle lift performance. The price of launch services is a less important factor; commercial buyers are willing to pay high premiums for small improvements in these three areas.

Selection as an NSS-certified launch provider can affect these areas and a launch provider's ability to win commercial contracts.

---

we and Supplier A independently adjusted the FAA forecast to this smaller market, we have confidence that our adjustment is reasonably correct.

Reliability and Insurability

First, commercial buyers value the reliability of launch vehicles to ensure that their payloads make it into orbit. Buyers may insure their payloads to hedge against launch failures and other potential technical issues. Insurance may cover the replacement value of the payload but not the value of the potential revenue stream.[12] Although the replacement value of a payload is typically on the order of $100 million to $500 million, the potential revenue stream for space service providers can be on the order of billions of dollars per year, which makes the reliability of launch vehicles crucial to maintaining the firm's financial viability.

Selection as an NSS-certified launch provider may signal higher launch vehicle reliability because of the completion of the nonrecurring design validation in the NSS launch certification process. To validate the launch vehicle design, (1) the design must be qualified for reliability and sufficient margin, (2) manufacturing processes must be qualified to produce the hardware designs, and (3) test programs must be verified to meet technical qualifications. NSS launch certification is a rigorous process, and launch service providers can market the reliability of their launch vehicles based on successful completion.

Scheduling Flexibility

Second, commercial buyers value not only the availability of revenue streams but also having adequate control over their timing. In many cases, time to market can make or break business cases for space service providers. Thus, commercial buyers value slack in launch manifests (e.g., launch capacities of at least 150 percent of manifests) to ensure that launch vehicles are available when satellites are ready for launch. The ability to open new revenue streams on time allows space service providers to compete more effectively in their markets.

Selection as an NSS launch provider can decrease scheduling flexibility for commercial buyers because of prioritization of NSS launches. The extent to which a commercial buyer's launch flexibility is affected will be dependent on the percentage of NSS launches on the provider's manifest.

Launch Vehicle Lift Performance

Finally, commercial buyers value the lift capabilities of launch vehicles to properly position their payloads in orbit. If payloads cannot be launched close to their final orbits, both time and fuel must be consumed to maneuver into the proper orbit. For satellites with liquid propulsion systems, the fuel required to change orbit can reduce the service lifetime of the satellite and significantly decrease lifetime revenue. For satellites with

---

[12] For an overview of the launch insurance industry, see Jeff Foust, "Do Smallsats Even Need Insurance?" *Space-News*, October 2018.

electric propulsion systems, the time required to change orbits can increase the time to market by four to six months.[13]

Selection as an NSS launch provider requires launch vehicles designed for NSS payloads, which require large masses to be launched into high orbits. Therefore, NSS-certified launch vehicles may not be optimally sized for the commercial market. However, an ability to launch multiple payloads on a single launch vehicle ameliorates this risk.[14]

Overall, commercial buyers of heavy lift launch highly value reliable launch services that can position their payloads in orbit when and where they are needed and are willing to pay high premiums to ensure their launch requirements are properly met. Although NSS-certified launch providers may have reliable launch vehicles that are also available to commercial buyers, these vehicles do not always have lift capabilities and scheduling flexibility well suited to commercial buyers.

Factors Launch Service Suppliers Consider When Assessing Competition in the Market

To understand competition in the heavy lift launch market, we also asked launch service providers to identify the factors they take into account when evaluating their fellow competitors. The factors they consider are the degree of government support, diversification in the launch markets served, and the competitor's overall commitment to the heavy lift launch market.

Government Support
First, launch service providers with strong governmental support have greater resources and flexibility to make the investments needed to effectively compete in the heavy lift launch market. Government support can be provided in many forms, including research and development funding to develop new launch vehicles, capital investments to build and sustain manufacturing capacity and/or launch infrastructure, or other funds to maintain technical expertise. Government support in the form of launch contracts can also help launch providers remain viable during market downturns or after a launch failure while confidence is being rebuilt. Russian government support for Proton after a string of launch failures from 2006 to 2016 may have been crucial for its survival (for more information, see Appendix C).

---

[13] Tereza Pultarova, "Largest All-Electric Satellite to Date Completes Orbit Raising in Record Time," *Space-News*, October 2017.

[14] "ULA Team Launches USAF's AFPSC-11 Multi-Payload Mission," *Air Force Technology*, April 2018.

Diversity of Launch Markets Served

Second, launch service providers with diversified portfolios of launch vehicles are less dependent on a single market and should be able to compete more effectively in the launch market overall. If the small and medium lift launch markets are robust, diversified launch providers should also be able to compete in the heavy lift launch market with smaller shares than other providers and have greater resiliency during market downturns. Similarly, commitments to customers in one market should not affect the ability of a diversified launch provider to serve customers in a different market and thus could allow better provisioning of multiple customers.

Commitment to the Market

Finally, launch service providers with strong commitments to the heavy lift launch market may continue to compete despite prolonged periods of unprofitability. Strong commitments may be derived from corporate goals to better compete in other aspects of their business or to maintain healthy relationships with important customers. Corporate commitment can also be derived from nonfinancial goals, such as returning to the moon or exploring Mars, or from the goals of furthering human exploration and the experience of space. Regardless of the reason, strong commitments can lead to unusual market dynamics not commonly observed in more robust markets.

CHAPTER FOUR

Strategic Choices Faced by U.S. Heavy Lift Launch Providers

U.S. firms must make strategic choices regarding how to effectively compete in the launch markets available to them. In this chapter, we discuss four potential strategies for competing in the various launch markets, based on the historical choices of U.S. and international firms, and how these strategies overlap in regard to the markets pursued (see Figure 4.1). The number of U.S. firms that can be supported by the demand in the launch markets will be dependent on the strategic choices of these firms.

To set up this discussion, we will first make a few observations about the relative sizes of the launch markets summarized in Table 4.1. First, the addressable commercial market exists because national policies globally allow the market to exist. Thus, other nations will either make the investments necessary to claim their share of the commercial market or will withdraw their demand, moving more launches from the addressable to nonaddressable side of the ledger. For this reason, U.S. firms are unlikely to be allowed to sustain greater than 50 percent of the total addressable commercial market (i.e., historically, ten launches a year) for any significant period; and, in fact, market share by nation has rarely been as lopsided as it is today.[1] Moreover, in Chapter Six, we argue that the long-term total share available to U.S. firms of the commercial addressable market is likely to be in the range of four to seven launches a year. In contrast, the U.S. NSS-certified launch market forecast is stable at seven to nine launches a year for the futures we analyzed. We did not develop a forecast for the demand in the other U.S. nonaddressable market, which consists primarily of ISS resupply, NASA launches, and self-provisioning. Historically, it has averaged three to five launches a year, but it is highly bursty.[2,3] With those relative sizes in mind, we now turn our attention to the

---

[1] The last time market share by nation was this lopsided was when Atlas and Delta dominated the market before the rise of Arianespace. In the 1998–2013 period, U.S. firms had a very small share of the global addressable market. Appendix C has additional perspectives on historical lessons from the heavy lift launch market.

[2] *Bursty* means to occur at intervals in short sudden episodes or groups.

[3] NASA's launch demand for non-ISS–related missions is highly variable and is affected by shifting national priorities and budgetary constraints. The National Research Council's 2014 study rates the cost and schedule risk of NASA's heavy lift and planetary ascent vehicles as High (see National Research Council, *Pathways to Exploration: Rationales and Approaches for a U.S. Program of Human Space Exploration*, Washington, D.C.: National

balance of this chapter's discussion on strategic choices facing U.S. firms that seek to position themselves in the heavy lift launch market.

Figure 4.1
Strategic Choices of Heavy Lift Launch Providers

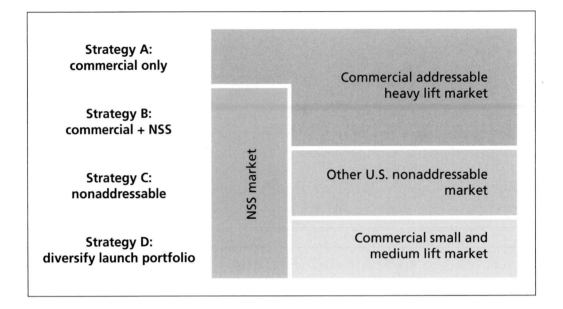

Table 4.1
Demand for U.S. Launch Services by Launch Market

Launch Market	Demand	
	Near Term	Long Term
Commercial	7–15	4–7
NSS	7–9	
Other nonaddressable	3–5	
Total	17–29	14–21

SOURCE: RAND analysis.
NOTE: This table's count of nonaddressable demand does not include self-generated demand.

Academies Press, 2014). Events since have affirmed that assessment. Although the current administration has ambitious plans to return to the moon and/or Mars, funding for the moon shot is estimated to cost between $20 billion to $30 billion (see Marina Koren, "The Fraught Effort to Return to the Moon," *The Atlantic*, July 17, 2019). The current fiscal year 2020 NASA budget coming out of the House of Representatives does not appear supportive of this effort (see Jeff Foust, "House Appropriators Take a Pass on NASA Budget Amendment," *Space-News*, May 16, 2019).

Strategy A: Specialize in Commercial Heavy Lift Launch Only

Under this strategy, U.S. firms could choose to focus solely on the addressable commercial heavy lift launch market. In this case, a firm would not pursue NSS launch certification and would size its launch capacity to the commercial market—diminishing the potential supply of NSS-certified launch vehicles. Serving only the commercial market would allow a firm to more easily maintain the launch schedule flexibility desired by commercial buyers and would enable it to compete more effectively in the commercial market. However, at our forecasted level of demand, a firm's ability to sustain itself using this strategy may be highly dependent on its commitment to the commercial market. To date, no U.S. firm has competed solely in the commercial market. Sea Launch made this strategic choice but was unable to compete effectively after a launch failure in 2007 and was forced to file for bankruptcy in 2009 (for more information, see Appendix C).

Strategy B: Specialize in Commercial Heavy Lift Launch but Maintain Availability for NSS Launch

With this strategy, U.S. firms could also choose to focus on commercial heavy lift launch but maintain availability for NSS launches on their manifests. In this case, a firm would need to size its launch capacity for both commercial and NSS launches and maintain enough schedule flexibility for commercial buyers despite prioritizing NSS launches. As discussed previously, commercial buyers expressed a preference for launch capacities greater than 150 percent of manifests to ensure schedule flexibility. This implies a need to keep NSS launches at a relatively small percentage of the total manifest. For example, if a firm accepts one NSS launch a year and wants to capture seven to ten commercial launch contracts, it would size for a launch capacity of 12 to 17 launches a year. If it accepts three NSS launches, it would size for a launch capacity of 15 to 20. This spare capacity would increase firms' overhead. Therefore, under Strategy B, the USAF would be required to pay a premium for NSS launch services above the firm's NSS planning number to compensate for the potential negative impacts on the firm's ability to compete in the commercial market.

At the levels of commercial demand forecast, this strategy appears viable only if a single U.S. firm captures all the share available to U.S. firms from the addressable market. **Therefore, the demand in the addressable commercial market is likely to support at most one U.S. firm choosing Strategy A or B.**

Strategy C: Specialize in Nonaddressable Heavy Lift Launch

Using this strategy, U.S. firms could focus on the nonaddressable heavy lift launch market instead of competing in the addressable commercial market. The total U.S. nonaddressable demand currently averages 12 launches a year, with NSS launches making up the majority of the demand. A firm using Strategy C should not expect to capture an appreciable number of commercial launch contracts, because sizing its capacity to be attractive to commercial buyers may be financially ruinous. For example, if the firm has six NSS launches and wants to be able to attract two commercial buyers, it would size its capacity to 12. When compared with simply sizing to build the six NSS launches, this is a significant increase in overhead. However, a firm using this strategy may also have access to NASA missions or self-provisioned demand. Of these, the self-provisioned demand may be the most attractive option for the firm to pursue, but only if it does not need scheduling flexibility for its own launches.

If firms do not have access to self-provisioned demand, it is likely that market demand can support only one Strategy A/B and one Strategy C provider. If no U.S. firm pursues Strategy A or B, the market could support two U.S. firms each with seven to ten launches a year from the total addressable and nonaddressable markets. These ABC firms will have a higher overhead than a purely Strategy C or B firm. The USAF should expect to pay a higher premium for priority on the manifest under this scenario.

If firms have access to self-provisioned demand, the market may be able to support a third U.S. launch provider. The premium the USAF would need to pay to obtain priority on the launch manifest will be dependent on the firms' ability/willingness to prioritize NSS launches over their own needs.

Strategy D: Diversify Launch Vehicle Portfolio

Finally, U.S. firms could pursue a strategy whereby they diversify their launch vehicle portfolios and compete in the small, medium and heavy lift launch markets. In this strategy, a firm could focus on the commercial markets for small and medium lift launch and prioritize NSS missions for heavy lift launch without jeopardizing schedule flexibility for commercial buyers. **The health of the small and medium lift launch markets, although not the focus of this study, may be supportive of a third U.S. firm choosing Strategy D.** Because a Strategy D firm does not need to carry the additional overhead capacity to attract commercial buyers in the heavy lift launch market, the USAF could expect to pay prices similar to those charged by a Strategy C firm. Diversification is also likely to make a Strategy D firm less vulnerable to downturns in any single segment of the launch market.

Evidence of Strategic Choices in the Heavy Lift Launch Market

In our review of historical launch manifests, we found evidence suggesting that the above strategic choices are indeed pursued in the market. The historical launch manifests for SpaceX, ULA, and Northrop Grumman, each suggestive of the strategies they pursued over the past decade, are shown in Figure 4.2. We strongly caution that the USAF not assume these firms will continue to pursue these strategies in the future. These firms might bring new capabilities to the market that necessitate an adjustment of their strategic choices. They might also change their strategic choices in response to other firms' choices. We also found historical evidence of foreign firms pursuing these different strategies (see Appendix C).

The Number of U.S. Firms the Launch Market Can Support Is Dependent on Firms' Strategic Choices

To summarize, if U.S. launch service providers do not have access to self-provisioned demand, it is likely that the total demand in the market can support at most: one Strategy A/B and one Strategy C provider; or two firms each with seven to ten launches a year from the total addressable and nonaddressable markets (i.e., Strategy ABC)

The market may be able to support a third U.S. launch service provider if firms have access to self-provisioned demand or if firms diversify to support the small and medium lift markets.

It is on the basis of the above analysis that we make our first recommendation: **The USAF should make prudent preparations for a future with only two U.S. providers of NSS-certified heavy lift launch, at least one of which may have little support from the commercial marketplace.** As noted earlier in this report, the USAF should expect to pay higher premiums for prioritized launch in the case of two ABC providers as opposed to one A/B and one C provider. Higher premiums for prioritized launches could also be expected in the case in which a third launch provider has access to self-provisioned demand versus a provider that has diversified to support the small and medium lift markets. Price, however, should be a secondary consideration and the USAF should not try to drive U.S. firms' strategic choices in an attempt to reduce costs. Firms must be free to adapt their strategies to changing market conditions.

In a monopsony, the dominant buyer can force providers to be "price takers." As the buyer with the greatest demand for heavy lift launch from U.S. providers, the USAF may have monopsony power in this market. To avoid a future in which providers sell to the USAF at a loss and then a need to consolidate (which was seen prior to the ULA merger, as described in more detail in Appendix C), the USAF must carefully balance its accountability to sustain a healthy launch market and its accountability to reduce the cost of NSS launches. A cost reduction obtained by wielding monopsony power is not sustainable.

Figure 4.2
Historical Launch Manifests and Strategic Orientations of SpaceX, ULA, and Northrop Grumman, 2007–2018

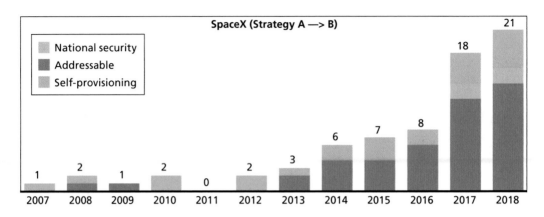

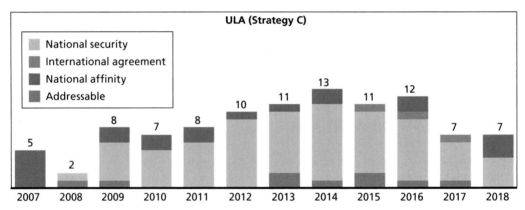

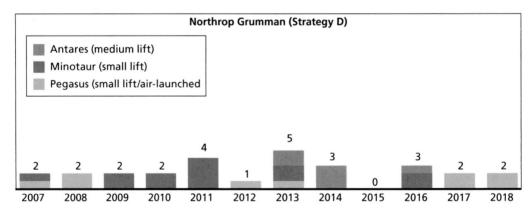

SOURCE: RAND analysis of heavy lift launch data set, 2007–2018 (Zak, 2019b; Union of Concerned Scientists, 2019; U.S. Department of Transportation, Bureau of Transportation Statistics, 2017; World Bank, undated; World Bank, 2018).

CHAPTER FIVE

Assessing the Impact of USAF NSS Launch Acquisition Decisions on the NSS Market

As noted earlier, the USAF is currently competing Phase 2 of its incremental approach to enhance competition in the NSS-certified launch market. Considerable concerns have been raised regarding acquisition decisions for this procurement. These concerns led the USAF to ask us to perform an independent analysis of the heavy lift launch market to assess the impact its near-term decisions might have on domestic launch service providers. [1]

In this chapter, we discuss how those decisions affect the USAF's ability to meet NSS launch demand using U.S. launch service providers. In the next chapter, we discuss how those decisions may affect the USAF's ability to sustain two or more U.S. launch service providers over the next ten years.

For review, the near-term USAF acquisition decisions we considered were

- the decision not to procure additional Atlas V launch vehicles prior to the 2022 congressionally mandated prohibition
- the ongoing Phase 2 procurement plan for NSS launches, specifically the decision to award contracts to two versus three launch service providers
- the decision to terminate research and development OTAs for providers not awarded a launch service contract under Phase 2.

To understand the impact of the above decisions on the USAF's ability to meet NSS launch demand, we ran Monte Carlo simulations of the range of futures the USAF could face over the next ten years. All cases assumed the current USAF acquisition plan as a baseline. For each two- provider and three-provider scenario, we then randomly sampled values for delays in first launch dates, firm decisions regarding capacity sizing, and slips in NSS demand. This approach, described in more detail in Appendix B, allowed us to make probabilistic statements regarding the impact of USAF near-term acquisition decisions.

---

[1]   We were tasked to identify costs, risks, and impacts of these decisions. To mitigate the possibility that cost identification would affect the Phase 2 source selection activities, this report addresses costs at an abstract-level only and focuses primarily on risks and impacts.

Supply Uncertainties

The Falcon Heavy is well along the path to NSS certification. ERBs for the launch vehicle are ongoing at the time of this report's writing, and the Falcon Heavy recently completed a successful third launch as required for certification. All other new entrants face significant development risks, with the earliest launch forecast for the second quarter of 2021. Vulcan just completed CDR and OmegA is forecast to complete CDR in the third quarter of 2019. New Glenn recently completed PDR. The underlying engines that make space flight possible have traditionally been a high development risk item. Engine risks for the new entrants are intertwined because of Vulcan's selection of the BE-4 engines also used on New Glenn and the GEM-63 engines also used on OmegA. Therefore, for all of these entrants, we assume that first launch dates are highly uncertain and model that uncertainty using a uniform probability that the actual first launch dates will occur between one and four years of the current forecast.[2]

In analyzing historical data regarding accuracy of forecast for first launch dates, we found: [3,4]

- There is no evidence that a forecast more than one year in the future will be achieved.
- For predictions of "next year," there is some chance that it will indeed be the last year. Falcon Heavy was forecast "next year" for five years, Antares for four years, Delta IV upgrade for two years. For new developments, there is no evidence that the maturity of the launch vehicle correlates to the number of years that "next year" is forecast, although we hypothesize there may be a correlation with government support/oversight.

---

[2]   Appendix B includes results for excursions from the baseline analysis to explore the sensitivity of our analysis to these entwined engine development risks.

[3]   Sources for first launch accuracy analysis: GAO, *NASA: Commercial Partners Are Making Progress, but Face Aggressive Schedules to Demonstrate Critical Space Station Cargo Transport Capabilities*, GAO-09-618, Washington, D.C.: June 16, 2009; GAO, *NASA: Medium Launch Transition Strategy Leverages Ongoing Investments but Is Not Without Risk*, GAO-11-107, Washington, D.C., November 22, 2010. Sources for Falcon Heavy forecasts: SpaceX, 2011; de Selding, 2015; Foust, 2015; Wall, 2016; Henry, 2017. Sources for Antares forecasts: Chris Bergin, "SpaceX and Orbital Win Huge CRS Contract from NASA," *NASASpaceFlight.com*, December 23, 2008; NASA, "NASA Launch Services Manifest," July 1, 2011 (at this point Antares is still called the Taurus II); Chris Bergin, "Space Industry Giants Orbital Upbeat Ahead of Antares Debut," February 22, 2012a; Chris Bergin, "Orbital's Cygnus Debut Mission to the ISS Outlined," *NASASpaceFlight.com*, June 4, 2012b. Sources for Delta IV Upgrade forecasts: Tim Furniss, "Launchers Directory," *Flight International*, December 11–17, 1996; *GlobalSecurity.org*, "Evolved Expendable Launch Vehicle (EELV) Program Overview," webpage, November 20, 1997; Boeing, "U.S. Air Force Procures Boeing Delta IV Launches for EELV Program," press release, October 16, 1998; Vandenberg Air Force Base, "Evolved Expendable Launch Vehicle (EELV)," fact sheet, August 4, 2017.

[4]   A 2010 GAO report states that NASA analysis of accuracy of third launch forecasts is three-plus years, on average: "Vehicles include all configurations of the Delta IV, two configurations of Atlas V, Falcon 1, Pegasus, and Taurus" (see GAO, 2010).

- For predictions of "three to six months," there is high probability of first launch within eight to 12 months (i.e., about eight to 12 months out, one can tell the difference between the light at the end of the tunnel and the train rushing toward you).[5]

Based on the above historical data, we believe our model on the accuracy of the first launch date forecast is realistic.

Capacity forecasts for the supply each new entrant will bring to the market are also uncertain. We obtained estimates for the projected supply capacity of the Vulcan, OmegA, and New Glenn in interviews we separately conducted to assess future demand for U.S. launch infrastructure. SpaceX has indicated that the Falcon 9 and Heavy are steppingstones to the Super Heavy that SpaceX will leverage to compete in the commercial market.[6] However, SpaceX has not yet filed a letter of intent for NSS certification for the Super Heavy. Therefore, in our NSS market analysis, we project that SpaceX's supply capacity for NSS launch (consisting of the Falcon 9 and the Falcon Heavy) will be between six and 20. The upper end of our estimate is based on a May 2019 statement from SpaceX's CEO that the company is sizing for a commercial market of 18 to 21 for a total capacity of about 40.[7] Table 5.1 shows our assumptions regarding the projected launch capacity for the four providers seeking NSS certification.

For each provider, we assumed an initial buildup curve after the first launch estimated from Falcon 9 performance during 2007 to 2012. With USAF support, it may be possible to achieve a higher build up rate than that achieved for Falcon 9 while also achieving 100-percent launch success, but we have no evidence to support such a claim (Falcon 9 buildup was unprecedented).

We assume USAF-supported firms dedicate their entire new entrant capacity to NSS launches. Depending on the firms' strategic choices for positioning themselves in the commercial market, this may be an unrealistic assumption, making our analysis a bounding case.

If firms were to increase their currently planned capacity ranges, it is unlikely to change our results—the risk of insufficient supply is primarily driven by delays in first launch and initial build up. On balance, we believe our model of supply uncertainty is neither optimistic nor pessimistic.

---

[5]   Evidence includes: (1) SpaceX was quoted as stating in June 2009 that the Falcon 9's first launch would occur "no earlier than three months," whereas it actually launched in June 2010 (a 12-month gap); (2) in public statements in June 2012, the chief executive officer of Orbital Science Corp. maintained that the first launch of the Antares would be within "three months," whereas it actually launched in February 2013 (an eight-month gap); and (3) in public statements in June 2017, the SpaceX chief executive officer asserted that Falcon Heavy's first launch would be within "six months," whereas the vehicle actually launched in February 2018 (an eight-month gap) (see GAO, 2009).

[6]   Caleb Henry, "SpaceX Targets 2021 Commercial Starship Launch," *SpaceNews*, June 28, 2019b.

[7]   Gwynne Shotwell as quoted in Caleb Henry, "SpaceX to Launch 'Dozens' of Starlink Satellites Next Week, More to Follow," *SpaceNews*, May 8, 2019a.

Table 5.1
Projected Range of Supplier Launch Capacity for the NSS-Certified Launch Market

Yearly Launches	Vulcan	OmegA	New Glenn	Falcon 9 and Heavy
Minimum	6	6	12	6
Maximum	12	8	16	20

SOURCES: RAND colleagues Gary Mcleod and Ellen Pint supplied the Vulcan, OmegA, and New Glenn capacity ranges and assessments space launch locations, which they obtained from interviews with providers. Falcon capacity range is based on RAND analysis of public statements made by SpaceX leadership regarding the commercial space market and the role of the Falcon 9 and Heavy in future business plans.

Demand Uncertainties

Figure 5.1 shows the U.S. NSS launch history and the USAF's current demand forecast. Current launch contracts cover demand through 2021 and for a portion of the 2022 demand. Under Phase 2 of the USAF's incremental approach to increase competition in the NSS launch market, unassigned launches in 2022 and beyond will be procured from new entrants who have completed or are on track to complete nonrecurring NSS certification.

Demand forecasts are also subject to uncertainty. Our analysis of historical *commercial* launch forecasts suggest that accuracy falls off rapidly if they are to take place more than a year into the future. However, the USAF must forecast demand and award contracts for launch vehicles two to three years before the forecast launch date to accommodate congressional budget timelines and to allow sufficient time for rigorous satellite interface design and integration with the selected launch vehicle. Therefore, we examined past forecasts of NSS demand to develop an uncertainty range for the demand forecasts. We found that those demand forecasts, on average, in the near term are reasonably accurate. However, the demand does tend to slip further out into the future as payloads are delayed, as shown in Figure 5.2. Historically, zero to 30 percent bounds the range of year-to-year demand slips for near-term NSS launches (with the exception of the years immediately surrounding the ULA merger [2006–2008] when delays seem to have been exceptionally pronounced).[8]

In our analysis, we use a uniform distribution of zero to 30 percent of yearly NSS demand slips and assume the procured launch vehicle slides with its demand (this avoids double counting a demand slip once it has occurred). Therefore, our results may be optimistic in the later years, but we feel they are realistic for near-term NSS-certified launches.

---

[8]   This analysis used the demand forecast from Forrest McCartney, Peter A. Wilson, Lyle Bien, Thor Hogan, Leslie Lewis, Chet Whitehair, Delma Freeman, T. K. Mattingly, Robert Larned, David S. Ortiz, William A. Williams, Charles J. Bushman, and Jimmey Morrell, *National Security Space Launch Report*, Santa Monica, Calif.: RAND Corporation, MG-503-OSD, 2006, and compared it with actuals shown in Figure 5.1.

Figure 5.1
U.S. NSS Launch History and Demand Forecast, 2007

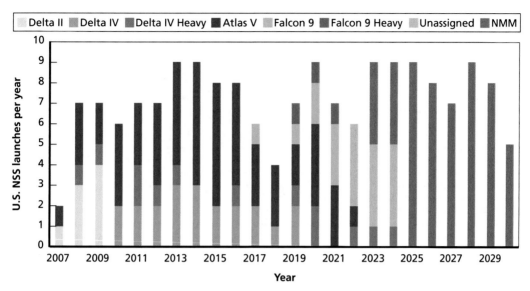

SOURCES: RAND analysis of data and informal notes provided by Space Missile Systems Center on ULA and SpaceX manifests for the 2019 National Mission Model (NMM) and the fiscal year 2020 President's Budget (Space and Missile Systems Center, "Launch Enterprise Systems Directorate Fiscal Year 2020 President's Budget," May 29, 2019).
NOTE: During the study period of performance, options were exercised for two Delta IV Heavies for launches in 2023 and 2024.

Figure 5.2
U.S. NSS Predicted Versus Actual Launches, 2005–2018

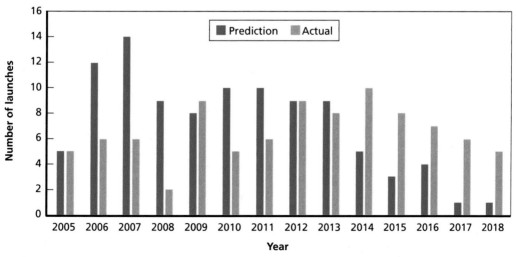

SOURCES: McCartney et al., 2006; RAND analysis of data and informal notes provided by Space Missile Systems Center on ULA and SpaceX manifests for the 2019 NMM and the fiscal year 2020 President's Budget (Space and Missile Systems Center, "Launch Enterprise Systems Directorate Fiscal Year 2020 President's Budget," May 29, 2019).

Significant Risk of Supply Shortages, Beginning in 2022

In many conceivable futures, the USAF's current acquisition plan is unlikely to provide a sufficient supply of NSS-certified launch vehicles to meet the projected demand. An exemplar plot of a typical result is shown in Figure 5.3 where we have graphed the average supply in excess of demand from 2021 to 2030. Also graphed are the +/–90-percent confidence levels. These confidence levels represent a determination that, given the assumptions of our model, we are 90-percent confident the actual supply of NSS-certified launch vehicles in excess of demand will be in this range. We found two distinct clusters of results in our analysis, one for when the USAF does not have access to supply from a provider who has currently achieved first launch as of 2020 (i.e., to legacy systems) and the other for when it does. Figure 5.4 shows the capacity in excess of demand for all futures where the USAF does or does not have priority access to legacy systems.[9]

These resultant supply shortages are not due to a USAF decision to select two versus three providers for the Phase 2 contract, although having access to a third provider could help mitigate the risk. The shortages are instead driven by the decision to begin filling NSS demand from among the Phase 2 providers starting in 2022 and the

Figure 5.3
Example of an Analysis Result Showing a Significant Supply Shortage Over Several Years

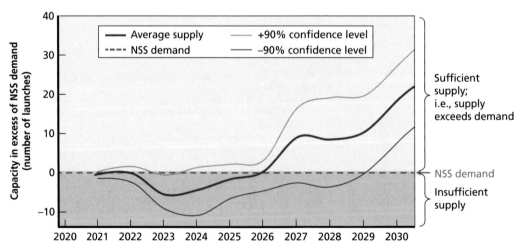

SOURCE: RAND analysis.

---

[9]   Figure 5.3 is a single two-provider scenario. In Figure 5.4, it has been averaged with other two- and three-provider scenarios where the USAF does not have access to legacy systems.

assumption we made regarding the achievable rate at which firms can build capacity for NSS launch. The Phase 2 model contract acknowledges this near-term supply risk and provides means for the USAF to acquire existing certified launch vehicles on a nonpriority basis in the first two years. As we will discuss later, these risk mitigation plans provide a means of supporting a third launch service provider without requiring the USAF to select that system as a primary provider for the Phase 2 contract.

Characterizing the Risk of Insufficient Supply on U.S. Warfighters

We created a framework to understand how insufficient supply of launch vehicles impacts end users of space enabled services, in this case the men and women of the U.S. defense and intelligence services. The USAF launches NSS payloads that are the global eyes, ears, networks, and timekeepers supporting the U.S.'s ability to project power across the globe. An insufficient supply of launch vehicles, if unmitigated, may result in delayed launch and unavailability of critical services, including those that enable U.S. precision strike capability, ballistic missile defense, and nuclear command and control networks. Any prolonged period of insufficient supply would affect mission readiness of the U.S. defense and intelligence services as shown in the logic model in Figure 5.5. [10]

Figure 5.4
Aggregated Analysis Results, 2020-2030, Sorted by Futures in Which USAF Does and Does Not Have Prioritized Access to a Legacy Launch System During Phase 2 Acquisition

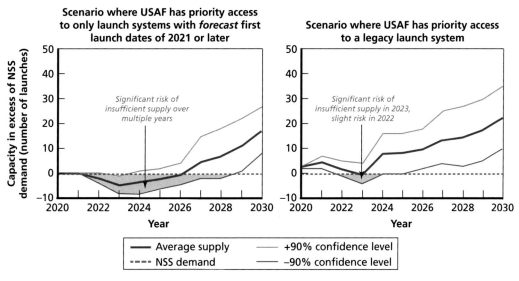

SOURCE: RAND analysis.

---

[10] Some DoD space-enabled services have excess capacity that provides resiliency if NSS payloads cannot be launched on schedule, while other services have no excess capacity and could be greatly affected.

Although the USAF has existing strategies and processes that reprioritize launch manifests to mitigate the loss of mission readiness caused by *demand* slips, different strategies and processes may be needed to compensate for *supply* shortages. To facilitate discussion of mitigations, we characterized the risk in terms of its possible consequence and likelihood as shown in Figure 5.6, which characterizes the impact of insufficient supply on mission readiness.

Consequence is determined as the percentage of planned service years in 2022 to 2025 that is lost because of supply shortage.[11] Risk of a particular acquisition strategy can then be graphed as the percentage of futures (likelihood) that result in the consequence or worse. We have chosen to use an approximation of logarithmic scales in assessing risk, which provides a better mapping of the color scale when computing consequence multiplied by probability to equal risk.[12]

We use this framework to characterize the risk of insufficient supply revealed in our Monte Carlo analysis in Figure 5.7, where four possible sets of futures are shown.

The highest risk set of futures, in the upper right hand corner of the Figure 5.7, is when the USAF does not have direct access to legacy systems (i.e., Atlas V, Delta IV Heavy, Falcon 9, and Falcon Heavy) under the Phase 2 contract. The second future set plotted is where the USAF does not have direct access to a legacy system but obtains nonprioritized access to one through options available through the Phase 1A contracts or the contingency clauses of the Phase 2 contract.[13] The third future plotted depicts the risk when the USAF selects any three providers, which could include selection of a legacy system, as primary providers under the Phase 2 contract. The final future

Figure 5.5
Risk of Insufficient Supply of Launch Vehicles, if Unmitigated, Leads to Loss of Mission Readiness for U.S. Defense and Intelligence Services

---

[11]   The total number of planned service years in this time period is computed based on the number of years that a new payload would be in service if launched on schedule—that is,

planned service launches = the number of 2022 launches in Figure 5.1 × 4 years

+ 2023 launches × 3 years × 2024 launches × 2 years + 2025 launches × 1 year.

[12]   Louis A. Cox, Jr., "What's Wrong with Risk Matrices?" *Risk Analysis*, Vol. 28, No. 2, April 2008, pp. 497–512; E.S. Levine, "Improving Risk Matrices: The Advantages of Logarithmically Scaled Axes," *Journal of Risk Research*, Vol. 15, No. 2, 2012.

[13]   This future is created using the following assumptions: In 2022 and 2023, the USAF successfully mitigates 85 percent of launch shortages, recovering all but six months of the service years for those launches. In 2024, the USAF successfully mitigates 50 percent of launch shortages, recovering one year of service for each of those launches. In 2025, the USAF does not attempt to mitigate the shortage, instead putting its efforts into executing a successful Phase 3 acquisition.

Figure 5.6
RAND Launch Mission Readiness Risk Framework

Consequence \ Likelihood	<2%	2–10%	10–20%	20–50%	50–100%
50–100% of planned service years lost (2022–2025)					
20–50% of planned service years lost					
10–20% of planned service years lost					
2–10% of planned service years lost					
Less than 2% of planned service years lost					

Low risk   High risk

set plotted depicts the USAF baseline plan to select two providers under the Phase 2 contract but mitigates the risk of supply shortages by exercising options under the Phase 1A contract and contingency clauses in the Phase 2 contract.[14]

As can be seen, for 20 percent of the worst futures, this latter strategy provides equal or better risk mitigation than selecting three primary providers under the Phase 2 contract. That is, the USAF baseline plan, with mitigations, is an effective means to support a third U.S. provider as recommended in this report. [15]

---

[14] The USAF has two remaining options under the Phase 1A contracts, one for an Atlas V for a fiscal year 2022 planned launch and one for a Falcon 9 for a 2024 planned launch.

[15] This is not to say that the baseline plan is the lowest acquisition risk—the USAF might need to pay higher prices for contingency access to legacy systems under the Phase 2 contract than they might have paid under a Phase 1A contract. Furthermore, there may be other risks that we did not analyze here (see "Factors the USAF Should Consider When Deciding Whether to Support Two Versus Three U.S. Launch Service Providers" in Chapter Six of this report).

Figure 5.7
Risk of Loss of Mission Readiness From Supply Shortages Under Four Possible Sets of Futures

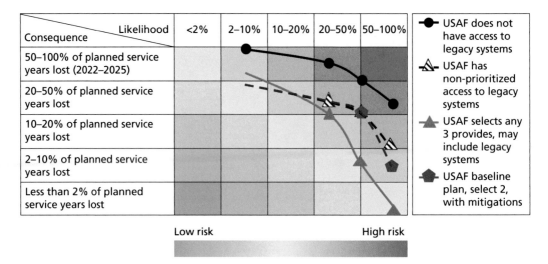

SOURCE: RAND analysis.
NOTE: Each point on this plot is read as "Under the assumptions documented in this report, the USAF may experience this consequence or worse in (likelihood) percent of futures"—that is, all futures in which the USAF does not have access to a legacy system are located to the upper right of the **solid** black line plotted. All futures in which the USAF mitigates this risk by obtaining nonprioritized access to a legacy launch system are located to the upper right of the **dashed** black line.

CHAPTER SIX

Assessing the Impact of USAF NSS Launch Acquisition Decisions on the Global Heavy Lift Launch Market

Earlier chapters in this report commented on the fact that the addressable market exists because national policies globally allow it to exist. Were any one nation to continually take a majority share of the market, we would expect other nations to adjust either their investments or policies to rectify this situation. In the case of Arianespace and the Russian Soyuz, there is every indication that the European Union and Russia are actively investing with the goal of retaking their "traditional" market share. Figure 3.2 in Chapter Three provided our perspective regarding the position of each competitor for heavy lift launch with respect to the global commercial market, using the IRL spectrum.

To understand the interactions between USAF near-term acquisition decisions and the dynamics of these resurgent international competitors, we expanded our Monte Carlo analysis beyond the U.S. NSS launch market. The expanded model has three key assumptions:

- Arianespace and Soyuz launch vehicles continue to compete in the market during the period studied (2020–2030) bringing new capacity to the market and improving their market share when they successfully introduce new vehicles into the market.

- SpaceX's Falcon 9 continues to compete in the market during the period studied (2020–2030). As the current market leader, it would be unreasonable to assume it would not. The model is agnostic to whether some or all of the Falcon 9 market share is replaced with Falcon Heavy. We assume that SpaceX rightsizes its capacity to the commercial market if not selected for NSS launch.

- Other U.S. firms that are not supported in their entries into the heavy lift launch market and foreign firms not currently in the market (Japan, China, and India) do not compete in the addressable market. This was a simplifying assumption that does not alter the conclusions of this report.[1]

---

[1] Additional U.S. or foreign entries would add to capacity and hasten market exits. In the real world, China currently participates in the global commercial market but at a low level. China's ability to participate has been hampered by the U.S. ITAR and Arms Export Control Act, but this effect may be fading. Japan does not compete overtly, but it is accepting launch contracts; India intends to compete but, based on our analysis of the market, has not yet crossed the chasm.

U.S. Firm Market Share Is Expected to Drop as Arianespace and Russia Field New Launch Vehicles Better Suited to Heavier Launch

One reason Arianespace is not currently more competitive in the heavy lift launch market is that the Ariane 5 was designed to carry two communications satellites to geosynchronous transfer orbit (i.e., to conduct dual manifest launches), allowing space services providers to share the cost of the ride to space. As satellites have gotten heavier, it has become harder to find a balancing lower weight payload to cost share with, especially since the two payloads must be placed in the same orbit.[2] This has removed a competitive advantage that made Arianespace a potent competitor in the market.

The Ariane 6 is designed in two models, one able to launch roughly 15 percent more weight than the Ariane 5 and the other slightly smaller. It will also have a reignitable engine, meaning that, for a dual manifest launch, the second satellite will no longer need to go to the same orbit as the first. Viasat recently announced that it would switch a current launch contract from the Ariane 5 to the Ariane 6. Eutelsat and OneWeb have also announced plans to launch on the Ariane 6. In the nonaddressable market, the French government and the European Commission have announced launches on the Ariane 6.[3] We expect that, with government support, Arianespace will regain market share when the Ariane 6 enters the market.

The case for Russia regaining market share is not as straightforward. Both the Proton and Soyuz launch vehicles have suffered failures because of workmanship issues. These failures correlate in time with periods of economic downturn in Russia and are not, perhaps, unexpected. However, we expect one or both vehicles to rebound with strong support from their national government and with the strong reputation of the Soyuz as the only viable provider of manned flight for crew transport to and from the ISS. For the analysis performed here, we assume this resurgence will be under the Soyuz banner, but the results are not dependent on which particular Russian vehicle reenters and/or regains market share.

We further assume a first launch date for the Ariane 6 of 2020 plus zero to three years and for the Soyuz (or other Russian vehicle) of 2022 plus one to four years. We assume the U.S. share of the addressable launch market will continue to be 30 percent to 67 percent before either of these new launch vehicles comes to market, that it drops to 30 percent to 50 percent after either come to market, and 20 percent to 30 percent after both. We further assume that the capacity of these firms grows when these new vehicles come to market, based on the historical quantities these firms previously supplied. For Arianespace, we assume capacity for the addressable commercial market

---

[2]   We performed an analysis of the weights of satellites currently in orbit and found that satellites above 2,000 kg have grown heavier with time while those below that level are growing much lighter. Even more telling is that the "medium" category of satellites has largely disappeared.

[3]   Viasat also has launch contracts with ULA and with SpaceX. Stephen Clark, "Viasat Swaps Ariane 5 Launch for New Ariane 6 Rocket," *Spaceflight Now*, June 17, 2019.

grows from the four to six range to the five to 11 range and for Russia, from the four to ten range to the eight to 14 range. For the U.S., the model assumes the capacity available for the commercial addressable market is that which is in excess of NSS demand. Demand for the addressable market analysis is assumed to be uniformly distributed between the bounds of our forecast shown previously in Figure 3.8 in Chapter Three.

This Monte Carlo analysis allows us to make probabilistic statements about the number of launches U.S. firms can expect to win in the addressable market given these assumptions. It is the basis on which we formulate our assessment regarding the number of U.S. firms the global heavy lift launch market can support. Results are shown in Figure 6.1. Given reasonable assumptions regarding Arianespace's and Russia's ability to regain market share, U.S. firms' share of the global market may drop to as few as four to seven launches a year by 2025.

Supporting Three Versus Two Providers May Reduce the Probability of Additional Nations Entering the Market but May Accelerate Eventual U.S. Firm Consolidation

The outcomes of our Monte Carlo simulations are sorted in Figure 6.2 by whether the USAF supports two versus three providers in their attempts to enter and/or continue in the commercial addressable market. Given our previous analysis indicated the size of the addressable market is unlikely to support more than two U.S. firms, the outcome of interest here may be *when* market consolidation occurs, not *if*. We have plot-

Figure 6.1
Projected Launches Won by U.S. Firms in the Global Addressable Market, 2020–2030

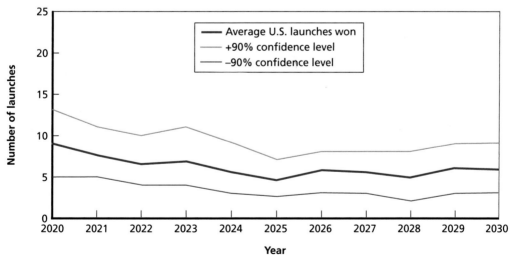

SOURCE: RAND analysis.

Figure 6.2
Global Commercial Launch Capacity in Scenarios in Which USAF Supports Two and Three U.S. Launch Providers, 2020–2030

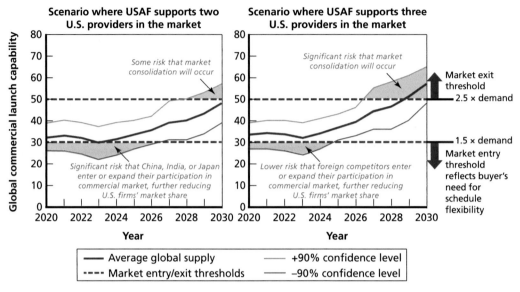

SOURCE: RAND analysis.

ted two lines on the charts: 1.5 times demand (1.5x) and 2.5 times demand (2.5x). At 1.5x demand, there is a greater risk of new nations entering or expanding their presence in the addressable market. This is the threshold at which commercial buyers we interviewed stated they would begin to seek out new launch providers in an attempt to increase overall scheduling flexibility. This may be the reason that Japan, despite not actively competing in this market, is currently accepting launch service contracts from traditional buyers. At 2.5x demand, we hypothesize there is a greater risk of market consolidation for nonstate supported launch providers (i.e., U.S. providers).[4] This is the classic case of "too many rockets, too few satellites." Either consolidation will occur or more nations will fence off more or greater portions of the demand in an effort to keep their domestic firms profitable.

As seen in Figure 6.2, limiting USAF support to two providers results in a two- to three-year delay in the eventual market consolidation. However, expanding to support three providers might discourage new nations from entering the market in the near term. In other words, supporting three U.S. providers may increase the probability of global supplier consolidation but decrease probability of additional foreign competition.

---

[4] The ULA merger provides some evidence for the market exit threshold. Boeing and Lockheed Martin had a combined 19 launches at their peak in 2003. In 2005, when the ULA merger was announced, Boeing and Lockheed Martin had a combined eight launches—making their combined capacity at least 2.4 times the number of launches they executed that year. For more information, see Appendix C.

Factors USAF Should Consider When Deciding Whether to Support Two Versus Three U.S. Launch Service Providers

Given that limiting USAF support to two providers results in only a two- to three-year delay in the eventual market consolidation and that it could encourage new nations to enter the market in the near term (and it may be difficult to dislodge them later), we recommend that the USAF continue to support as many entrants as it can afford to enter and/or continue in the heavy lift launch market. Note that this support need not be expensive: Sometimes it is only the removal of prohibitions that is needed.[5] However, as we saw in our discussion regarding the strategic choices that U.S. firms must make regarding how to position themselves in the markets, a firm that chooses to accept NSS launch contracts might need to incur additional overhead costs to demonstrate launch schedule flexibility versus when selling solely in the commercial market. Although we fully support the USAF's desire to pay only their fair share of the firm's overhead, a rethinking of what constitutes a fair share might better support U.S. firms in their attempts to compete in the commercial markets.

We also note that supporting three providers does not necessarily mean selecting three providers for Phase 2 acquisition. In fact, it may be in the U.S. firms' best interest to limit the number of NSS launch contracts to two, given our assessment that the total market supports at most either (1) one Strategy A/B and one Strategy C provider or (2) two U.S. firms each with seven to ten launches a year from the total addressable and nonaddressable markets (i.e., Strategy ABC). That said, it may be in the USAF's best near-term interest to select three providers to mitigate the risk of supply shortages in the 2022–2024 time frame. Alternatively, the USAF may elect to contract with a third provider only when necessary to mitigate supply shortages as is provided under the current contingency clauses of the Phase 2 contract. A careful balance of all competing interests is required.

The USAF has argued that supporting three launch service providers will impact its ability to provide full mission assurance and 100-percent mission success.[6] There are two components to this argument. The first is that spreading the limited number of NSS launches (seven to nine a year) over three providers does not provide enough launches to keep the requisite launch teams in a state of readiness. The 2006 *National Security Space Launch Report* states: "The number of launches needed to establish reli-

---

[5]   A nonmonetary means of supporting U.S. firms might be removing restrictions on use of surplus ICBM engines for commercial launches (see GAO, *Surplus Missile Motor Sale Price Drives Potential Effects on DOD and Commercial Launch Providers*, Washington D.C., August 2017b). Another possible nonmonetary means could be for Congress to make the 2022 ban on Russian-designed and Russian-manufactured engines contingent on the successful launch of alternative NSS-certified launch vehicles. If no launch service provider selected under Phase 2 has successfully launched by 2021, the ban could be delayed for a certain duration; or if only one launch service provider has successfully launched, the ban could be delayed a shorter duration.

[6]   Sandra Erwin, "Air Force to Continue to Push Back on Proposed Space Launch Legislation," *SpaceNews*, June 27, 2019.

ability is subjective, and different government studies have yielded different numbers. However, assessments are reasonably consistent and range between two to four launches per year to sustain confidence in reliability and resiliency while providing some program stability."[7] As long as U.S. firms do not pursue an NSS-only strategy, it appears that dividing the NSS demand between three providers need not have a negative impact on mission assurance.

The other component of the argument, however, concerns the human resources expended on contracting activities, nonrecurring certifications of launch vehicles, and on recurring certifications of individual launch vehicle/payload combinations. We will discuss each of these activities in turn.

- For contracting, although the USAF has a proven track record of being able to efficiently execute Phase 1A contract options, the resources needed to execute the contingency clauses of the Phase 2 contract are less certain. It is possible that a Phase 2 provider might elect not to take advantage of the opportunity to offer a legacy launch service in the event of a supply shortage, leaving the USAF to negotiate separate launch contracts for individual launches. If this is the norm and not the exception, contracting resources may be better utilized by selecting three Phase 2 providers. If it is the exception, then it may be reasonable to assume that selecting two providers minimizes USAF and industry resources expended on contracting efforts.
- Nonrecurring certification efforts are not related to whether a provider has a current NSS launch contract, but only on the provider's expressed intent to certify. Avoiding the nonrecurring certification costs of additional providers is only possible if the launch service provider withdraws from the NSS market. In response to our concerns regarding how a decision to select two vs. three launch providers for the Phase 2 contract would impact these resources, the USAF provided the following statement:

> The Launch Enterprise will fully support Phase 2 launch provider's development to ensure mission success. The Launch Enterprise will continue Non-Recurring Design Validation (NRDV) on a non-interference basis for unsuccessful Phase 2 offerors. The depth and timing of the support will depend on which offerors choose to continue their NRE and the Phase 3 acquisition strategy. The Phase 3 strategy will be based on market conditions, the initial results of the National Security Launch Architecture study, and other factors such as DoD Budget priorities, etc. The Phase 3 strategy will be coordinated for approval over the next 1-3 years and could include a development/investment phase (Author email exchange with Space and Missile Systems Center Launch Enterprises, January 31, 2020).

---

[7]   McCartney et al., 2006.

Therefore, we believe that selecting two launch service providers for Phase 2 might offer greater flexibility to the USAF in how it supports nonrecurring certifications efforts. However, we caution that it does not eliminate the need for those resources.

- Recurring NSS certifications are conducted on each combined payload and launch vehicle "stack." This certification process includes oversight of all launch vehicle/payload integration activities and consumes significant resources from the USAF, its federally funded research and development centers (FFRDC), the payload providers, and the launch service providers. Although standard interfaces have reduced the risk of payload and launch vehicle incompatibilities, the general practice is to identify the launch vehicle for a given payload approximately two years before launch to allow for these critical interface activities and verification efforts. To reap the full benefit of three providers in reducing the risk of insufficient supply, the USAF might need to certify schedule-critical payloads against all three providers' launch vehicles. The USAF provided an estimate of 25 to 40 labor-years of effort for recurring certification of legacy launch vehicles (depending on complexity of the analysis) and up to 80 labor-years for a first recurring certification of new entrants. This labor estimate covers only USAF and FFRDC effort; additional costs would be borne by the launch provider and the payload provider. Certainly, this would be prohibitive if applied to all payloads. More importantly, the resulting dilution of attention paid to each integration could indeed negatively impact mission assurance. Prudent selection of the number and types of payloads that need to be designed to be compatible with more than one launch vehicle can mitigate this risk but may reduce cost savings the USAF anticipates from having common interfaces.

In summary, while the USAF has an impressive record of NSS launch reliability, sustaining a near 100 percent success rate, there are legitimate concerns that the human resources needed to maintain that record may be stretched thin if the USAF were to select three launch providers for near term acquisitions.

CHAPTER SEVEN

Observations and Recommendations

In 2019, the USAF asked RAND researchers to independently analyze the heavy lift launch market to assess how decisions it might make in the near term could affect domestic launch providers.[1] The research team examined historical and projected levels of demand and supply in the global commercial and NSS launch markets to help the USAF gain clearer insight into the following issues:

- the number of U.S. launch service providers that the global heavy lift launch market can support
- the impact of near-term acquisition decisions on the USAF's ability to (1) meet NSS launch demand using U.S. launch service providers and (2) sustain two or more U.S. launch service providers over the next ten years.

To pursue this research, the research team (1) conducted a literature review, (2) examined historical launch data, (3) reviewed actor-specific launch data, and (4) carried out interviews with subject matter experts and industry stakeholders. The research team used data from these efforts to develop a forecast for the future heavy lift launch markets and constructed vignettes to illuminate strategies U.S. firms could use to position themselves in the marketplace. It also produced a series of Monte Carlo simulations to determine how near-term USAF acquisition decisions might play out over various possible futures.

Number of U.S. Launch Service Providers the Global Heavy Lift Launch Market Can Support

The research team found that the number of launches worldwide grew to 71 in 2018 from 47 in 1998. However, the research team also found that the portion over which

---

[1] The research team was tasked to identify costs, risks, and impacts of these decisions. To mitigate the possibility that cost identification would affect the Phase 2 source selection activities, this report addresses costs at an abstract level only and focuses primarily on risks and impacts.

launch providers compete—the so-called addressable share—remained steady during that period at an average of 20 launches a year. The addressable share represents only 35 percent of the total launch market today. The research team also found that SpaceX's Falcon 9 handles more than half of addressable launches today, with telecommunications companies being the dominant customers. We also found that the U.S. share of the market is unlikely to remain at its current greater than 50-percent level. Thus, the size of the global addressable market, the leading position that SpaceX has in this market, and the fact that the U.S. share is likely to shrink mean that the global heavy lift launch market is unlikely to support more than one U.S. provider of launch services focused on commercial heavy lift.

Impact of Near-Term Acquisition Decisions

USAF Ability to Meet NSS Launch Demand Using U.S. Launch Service Providers

The research team found that, in many conceivable futures, the USAF's current acquisition plan is unlikely to provide sufficient supply of launch vehicles certified to carry U.S. NSS payloads in the 2022–2025 time frame. However, the team also found that the USAF can lower this risk of insufficient supply of NSS launch vehicles by using its existing options under the Phase 1A and Phase 2 contracts to obtain nonpriority access to legacy launch systems. Alternatively, the USAF could mitigate the risk by selecting a third provider under the Phase 2 contract. In either case, the USAF would be supporting a third provider as recommended in this report.

USAF Ability to Sustain Two or More U.S. Launch Service Providers Over the Next Ten Years

The commercial addressable market share held by U.S. firms is expected to drop as Arianespace and Russia field new launch vehicles that are better suited to heavier launch. Our Monte Carlo simulations suggest that the market share held by U.S. firms might drop to as few as four to seven launches a year as early as 2025, making it likely that NSS launches will be the dominant source of demand for U.S. heavy lift launch over the next decade. The research team also observed that if the USAF supports three U.S. launch service providers, rather than two, it might reduce the probability of additional foreign competitors—such as Japan or India—entering the commercial market; however, in doing so, the USAF may accelerate eventual market consolidations.

Risks Identified by Our Study

Over the course of the study, we identified three primary risks that the USAF should consider when making near-term launch acquisition decisions.

- The national security risk of not having assured access to space in times of need because of a shortage of NSS-certified launch vehicles. Although this report identifies and quantifies a near-term supply shortage, this risk might also arise if a certified provider is removed from the market temporarily because of a launch failure or permanently after a business failure or market consolidation. Therefore, this report explores the risk of market consolidation as a function of U.S. firms' business strategies. Mitigating this national security risk is our top priority in formulating our recommendations.
- The increased acquisition costs that the USAF could incur if the burden of maintaining two NSS launch providers cannot be shared with the commercial market. This has been the situation for the past two decades, changing only recently with SpaceX's certification for NSS launch. Although near-term USAF decisions might shape this risk in ways that are beneficial to both the U.S. government and U.S. launch service providers, it is unlikely to be avoided altogether.
- The technical risk of any individual NSS launch failure. Although the USAF has an impressive record of NSS launch reliability, sustaining a near 100 percent success rate, there are legitimate concerns that the resources needed to maintain that record may be stretched thin if the USAF were to select three launch providers for near-term acquisitions.

Mitigating each of these risks to an acceptable level may require more resources than the U.S. government currently allocates to the acquisition of national security launch services and the sustainment of assured access to space. Because choices and balance among these risks are such critical issues, we requested clarification from Space and Missile Systems Center's Launch Enterprise regarding the current prioritization of these risks:

> The Launch Enterprise's top priority is mission success. In addition, the guidance from Air Force and DoD leadership, the White House Staff, and Congress was to urgently end the use of the Russian RD-180. Accordingly, when developing our acquisition strategy, we prioritized ensuring mission success over the risk of only having a single provider for a short period of time (i.e., an Assured Access to Space Gap). Phase 2 involves transitioning our most critical payloads to new launch systems, ending the use of the Russian engine. We deliberately focus Phase 2 execution on mission assurance to sustain 100% mission success but will work to minimize the risk to Assured Access to Space to the maximum extent practicable.[2]

In light of the magnitude of the assured access risk we found in this study, we believe a larger policy conversation regarding resource allocation is needed. We encourage the U.S. Congress and the larger NSS community to engage in a meaningful

---

[2] Email communication with author, February 21, 2020.

dialog with the USAF regarding how to prioritize these launch-related risks within the larger context of all national security risks that the USAF must balance.

Recommendations for the USAF

Long-Term Strategy

The USAF should make prudent preparations for a future with only two U.S. providers of NSS-certified heavy lift launch, at least one of which may have little support from the commercial marketplace.

Near-Term Strategy

The USAF should continue to provide tailored support through 2023 to enable three U.S. providers to continue in or enter the heavy lift launch market, including the exercise of remaining options under the Phase 1A contracts and the possible use of contingency clauses in the Phase 2 contracts. The research team notes this approach lowers the risk of loss of assured access to space for NSS payloads and may decrease the likelihood of additional foreign competition in the global addressable market.

Finally, supporting three providers in the short run has longer-term benefits in that it provides time for U.S. firms to adapt and position themselves in the launch markets and allows market forces (not the USAF) to determine which firms are strongest, and thus survive, and which exit. We emphasize that supporting three providers does not necessarily mean selecting three providers for the NSS Phase 2 acquisition.

Data Sources

Our analysis and recommendations draw primarily from four categories of data: (1) literature review; (2) historical launch data; (3) actor-specific launch data; and (4) interviews with subject-matter experts and industry stakeholders.

Literature Review

To understand the space launch market, we reviewed the available literature on the history of heavy lift launch and government investments in it. Much of the historical information we gathered on the key players in the launch market is captured in Appendix C.

We also reviewed previous efforts to measure and forecast the launch market in the near and medium term, public statements of payload, satellite service, and launch service providers regarding their plans and/or past performance, and GAO and government studies regarding the impact of government actions on the launch market.

Various efforts have been made to forecast future demand for both the NSS-certified launch market and the global commercial launch market. For NSS launch, the congressionally mandated National Security Space Launch Requirements Panel[1] projected NSS launches between 2005 and 2020. Although this projection successfully captured the nature of launches over time, we observed consistent slips in the launch manifest. This observation defined the range of NSS demand slips used in our analysis. On the commercial side, the FAA publishes ten-year forecasts on commercial orbital launches through its *Annual Compendium of Commercial Space Transportation* report. Given the FAA's consistent overestimation of launch demand, we developed our own forecast using observations from historical launch data.

---

[1] McCartney et al., 2006.

Historical Launch Data

To identify historical trends in the addressable launch market, we gathered satellite and launch data from a variety of sources. Originally, we hoped to gain electronic access to the FAA's *Commercial Space Transportation Year in Review* and *Annual Compendium of Commercial Space Transportation* reports, which date back to 1994. In discussions with FAA personnel, it became apparent that our study (which lasted just 45 days) would be completed before the appropriate approvals could be obtained. The one exception to this was the Department of Transportation's "Table 1-39: Worldwide Commercial Space Launches," which tabulates FAA designated commercial launches by launch service provider from 1994 to 2018. Note that neither the FAA nor the Department of Transportation's "Table 1-39" have yet been updated to include 2018 launches. Given inaccessibility to the FAA's electronic information, we turned instead to RussianSpaceWeb.com's yearly table on the world's orbital launch attempts and the Union of Concerned Scientists' "UCS Satellite Database," both of which were available in an electronic format suitable for analysis.[2] Both the RussianSpaceWeb.com and FAA reports contain information at the launch-event level. The Union of Concerned Scientists' database contains information at the individual-satellite level and only contains information for satellites currently in orbit. Unfortunately, this meant we needed to look elsewhere for information regarding satellites that failed to reach orbit or that have failed in orbit. We filled these gaps with information from payload and space service and launch service provider websites. For data reported at the satellite level, satellites were grouped into a single launch–event based on their shared launch dates and sites.

Data from our two electronic sources were merged together at the launch-event level, producing a data set of 677 heavy lift launches between 2007 and 2018. We cross-referenced these launch events with data from the FAA's *Annual Compendia of Commercial Space Transportation* and the U.S. Department of Transportation.[3] Inconsistencies and missing data were flagged and verified based on open-source reporting from payload, space service, and launch service providers as shown in Figure A.1.

Within our data set, heavy lift launches represent 68 percent of all global launches. We define heavy lift launches on a sliding scale where a launch vehicle capable of "heavy lift" in 1998, such as the Molniya-M, might only be considered a medium lift vehicle had it entered service in 2018. Our data set maintains the classification of such launch vehicles as heavy lift until they exited service. The full list of vehicles defined as "heavy lift" is defined in the RAND Global Launch Data Set.

---

[2]   Union of Concerned Scientists, 2019; Zak, 2019b; Gunter's Space Page, homepage, undated; SpaceFlight-Now.com, homepage, undated. We also used Gunter's Space Page and SpaceFlightNow.com to cross check our data set. When we could not determine a consensus from these data sources, we used the FAA's Year in Review as our definitive source.

[3]   FAA, 2013; FAA, 2018; U.S. Department of Transportation, Bureau of Transportation Statistics, 2017.

Figure A.1
Sources of Data Used to Create Global Heavy Lift Launch Data Set, 2007–2018

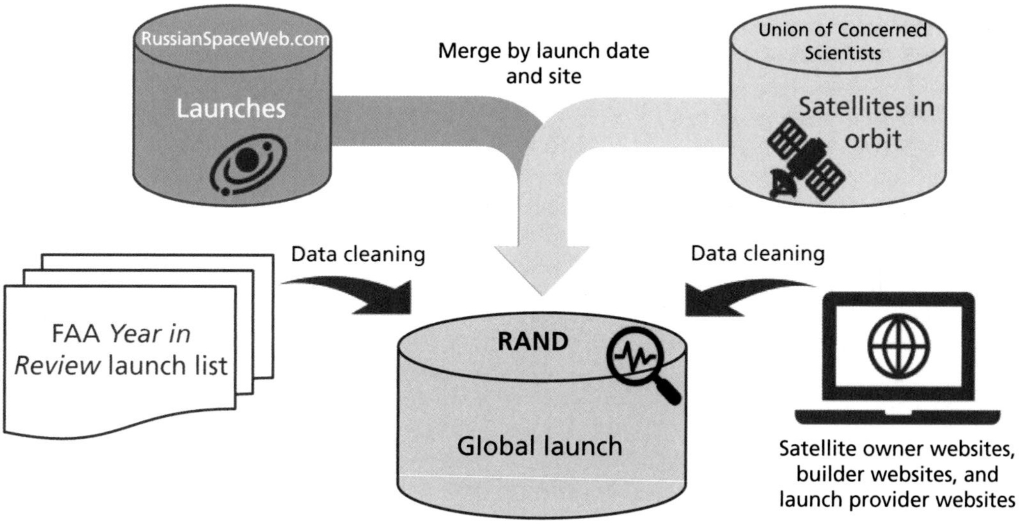

Of the 677 heavy lift launches, we coded 235 as "addressable" by U.S. launch providers based on their ability to compete for such contracts. Excluded from the "addressable" category are national security launches, launches subject to international agreements, launches of national affinity, and self-provisioning launches.

To understand addressable heavy lift launch services, we analyzed the purpose, class of orbit, and country of the operating entity (operator country) for the primary payload of each launch. We defined *primary payload* as the largest mass satellite. We then added World Bank data on national income groups and geographical region based on country characteristics of the operator country at the time of launch. Income group classifications for the 2007–2017 period followed the World Bank "GNI Per Capita Operational Guidelines & Analytical Classifications (FY19)," while classifications for 2018 were interpolated using 2017 data.[4] Finally, satellite data from the Union of Concerned Scientists were used to calculate the total launch mass of each addressable payload. A full list of variables and their sources is shown in Table A.1 and Figure A.2 provides a pictorial view of the data set.

Actor-Specific Launch Data

In addition to industrywide historical data, the forecast considers different eventualities at the actor level that may affect the supply or demand of launch vehicles. On the supply side, this includes the annual capacity sizing decisions of U.S. firms, as well as

---

[4]   World Bank, "World Bank Country and Lending Gap," webpage, undated.

Table A.1
Sources for the RAND Heavy Lift Launch Data Set

Source	Level of Observation	Variables
RussianSpaceWeb.com	Launch event	Launch vehicle
		Launch site
		Date of launch
		Launch result
		Payload
Union of Concerned Scientists' satellite database	Satellite in orbit	Operator country
		Total launch mass (kg)
		Total dry mass (kg)
FAA *Year in Review* tables	Launch event	Class of orbit
		Commercial classification
United Nations/World Bank	Country	Country code
		Region
		Income group
RAND analysis	Launch event	Satellite purpose
		Addressability

the entrance and exit of launch vehicles in the market. Our analysis focused on current and future launch vehicles capable of heavy lift launch. Based on interviews and open-source information, we developed an estimate of the future launch capacities for each provider. For ULA, Blue Origin, and NGIS, we used capacity ranges obtained by a separate RAND team evaluating the capacity of U.S. launch infrastructure. SpaceX capacity ranges were estimated from public statements by company leadership.

The demand for launch services can be divided into the addressable and nonaddressable markets. For addressable demand, our forecast relies heavily on trends in the historical launch data. However, this prediction also factors in possible macro shifts in launch service usage by major satellite operators (such as Intelsat, SES, Eutelsat, Echostar, Telesat, JCSAT, and Inmarsat) based on their announced plans for satellite launch. For nonaddressable demand, we projected the U.S. historical demand for NSS launches from the National Mission Model (2019) and the fiscal year 2020 President's Budget, with backing from the Launch Information Support Network database

(2019).[5] We further analyzed the NSS launch procurement process to identify options for the Space and Missile Systems Center to guarantee assured access to space.

Not-for-Attribution Interviews

Our final source of data for this analysis was a series of unstructured interviews conducted on a not-for-attribution basis with both foreign and domestic payload providers, space service providers (and their insurers), and launch service providers. Specifically, we probed for factors that buyers found important in selecting launch service providers and that providers used in evaluating their competition in the launch markets. These interviews were frank, open, and far ranging. All freely shared with us their perspective regarding future directions in the launch market and our analysis is infinitely richer for those conversations. Unfortunately, we were not able to interview the payload operators in lower- and middle-income nations that our analysis identified as a source of growth.

---

[5] RAND analysis of data and informal notes provided by Space Missile Systems Center on ULA and SpaceX manifests for the 2019 National Mission Model and the fiscal year 2020 President's Budget (Space and Missile Systems Center, 2019).

Figure A.2
Pictorial Representation of the Sources of Data in our Heavy Lift Launch Data Set

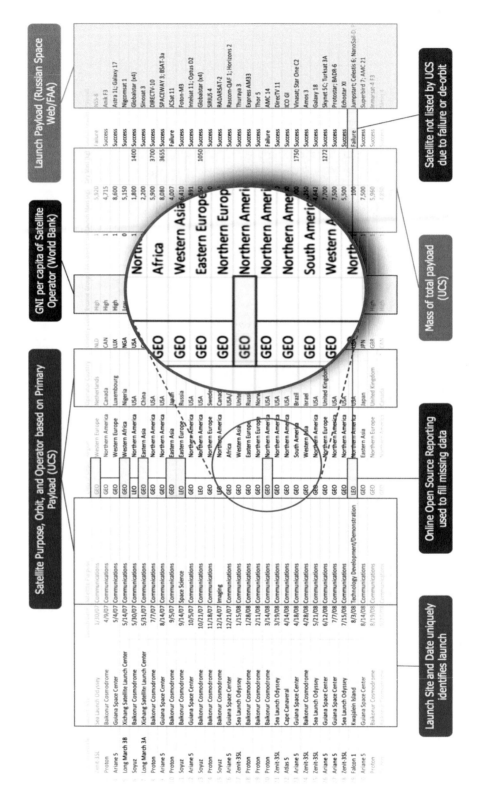

Methodology

Overview

To support this analysis, we employed a variety of Monte Carlo simulations. These simulations, in a broad sense, aim to estimate the expected outcomes and associated confidence intervals of various launch supply and demand scenarios by randomly drawing relevant parameters from presumed probability distributions, computing the outcome of the scenario, and repeating this process enough times to accurately approximate the distribution of outcome variables.

Although there is no standard, precise definition of Monte Carlo methods, they tend to follow this general procedure: (1) define input domains; (2) randomly generate inputs from their respective domains; (3) perform some calculation(s) on these inputs to generate outputs; (4) repeat steps 1 through 3 many times; and, lastly, (5) amalgamate the outcomes of all simulations in some fashion. It is worth taking a step back to discuss how these methods can be used in general and why they are particularly well-suited for the analysis at hand. Consider the following example: What is the probability that the sum of two die throws equal eight? One way to solve this problem—the analytical approach—is to write down all possible outcomes of two die throws, compute the sum in each case, and take the number of outcomes where the sum equals 8 and divide this number by the total number of possible outcomes. Alternatively, one could throw two die many times, calculate the number of times the sum equaled 8, and divide that number by the total number of times two die were thrown to approximate the desired probability. As the number of die throws increases, this approximation becomes arbitrarily close to the "true" probability obtained by solving this problem analytically—this is the idea behind Monte Carlo methods. Moreover, there is no need to physically throw dice thousands of times and manually compute their sum to approximate this probability; computers can do this in a matter of seconds by randomly drawing an integer between 1 and 6 to simulate each throw of a die.

In this example, it is unclear whether constructing and performing this simulation exercise would save time relative to solving the problem by hand, because the set of possible outcomes of two die throws is relatively small. But if the question is instead to calculate the probability that the sum of 10 die throws equals 36, or that the sum of 100 die throws equals X, a Monte Carlo approach becomes more attractive. As the complexity of these types of questions increases, analytical methods—while still possible—are impractical. Furthermore, there are certain types of questions that are impossible to answer analytically even if one was determined to try. Our space launch supply-and-demand analyses draw parallels to estimating the probabilities of an assortment of outcomes from throwing a large number of dice. In principle, given the input uncertainties of supply and demand, we could manually identify each possible launch outcome in each year under each NSS launch provider scenario and then calculate the probability of each occurring. In practice, doing so is unrealistic.

Additionally, in our context there are further benefits to using Monte Carlo methods beyond computational ease. One is that our direct simulation outputs align closely with our research questions, providing a measure of launch supply, launch demand and net capacity—capacity minus demand—over time. The simulations naturally produce many values for each of these variables over time, making it straightforward to plot items such as mean (average) capacity and statistical ranges (maximum minus minimum values) over time. Another related benefit is the ability to easily evaluate the likelihood of myriad outcomes—or ranges of outcomes—occurring. By binning outcomes using histograms, we can determine the relative likelihood of each outcome, given the assumptions the model is built on. In this vein, Monte Carlo methods allow researchers and analysts to quantify the sensitivity of outcomes to the uncertainty embedded in each individual input parameter—an invaluable tool when conducting any sort of risk assessment. Whereas expert judgement can be used to elicit the most likely, best-case and worst-case scenarios to expect, those estimates are inherently accompanied by uncertainty. Monte Carlo methods allow researchers and analysts to quantify this uncertainty and its impact on outcomes. Figure B.1 illustrates how our data sources, modeling assumptions and the model itself lead to outcomes.

We conducted two sets of simulations—one aimed at the domestic NSS launch market and another to explore the global commercial launch market. Our global commercial launch market simulations can be further divided into those focused on examining this market from the perspective of U.S. firms and those concerning the market in aggregate. Further details are provided below; here we restrict attention to describing the broad contours of each set of simulations. Domestic NSS launch market analysis is designed to understand the risks associated with the U.S. government's ability to launch forecasted NSS payloads on schedule under each possible launch provider scenario. This necessitated simulating both launch supply scenarios (i.e., the capacity supplied by each launch provider) and launch demand scenarios (i.e., the number of payloads the U.S. government wishes to launch). Launch supply in any particular year is given by the sum of launch capacity across all certified NSS providers. In any given year, launch capacity for a specific provider is a product of the provider's randomly drawn maximum yearly launch capacity (which is taken from a prespecified range) and where it stands in its buildup process as it progresses from first launch toward maxi-

Figure B.1
Data Sources, Modeling Assumptions, Simulations and Outputs of Monte Carlo Analyses

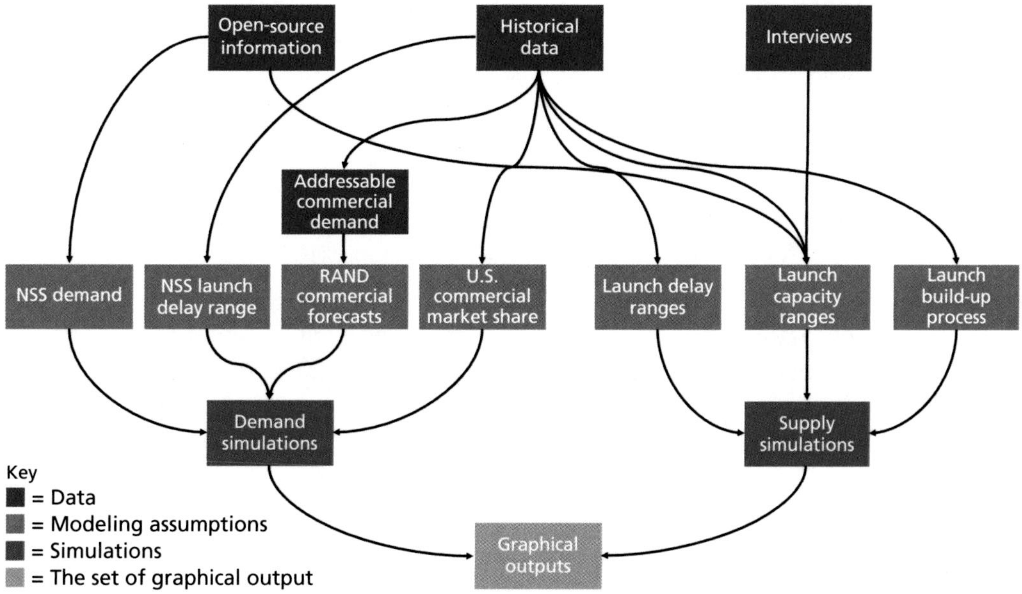

SOURCE: RAND analysis.

mum annual output. On the demand side, we used the NSS launch demand forecast of Figure 5.1 in Chapter Five and, each year, allow a variable fraction of launches to experience a one-year launch delay.

A typical plot resulting from our Monte Carlo analysis of the NSS launch market depicts, by year, the range of outcomes across all simulation iterations of NSS launch capacity less NSS launch demand under various provider selection scenarios. In some cases, these plots are converted to instead display average net capacity across various different launch provider scenarios, along with associated 90-percent confidence intervals. Upon constructing these plots, we conduct a more detailed analysis for each year between 2022 and 2025. These are years where net capacity ranges venture into negative values—i.e., where projected supply does not meet launch demand. The more detailed analysis uses histograms to convey the breakdown of all simulation outcomes for the years in question. Histograms allow us to ascertain the percentage of outcomes resulting in negative net launch capacity and hence provide deeper insight into prospective risks to the U.S. government and let us directly map that risk onto the risk framework we used in our analysis.

The nuts and bolts underlying the heavy lift addressable commercial market simulations are similar to those behind their NSS counterparts, although these simulations serve a different purpose. Recall that we undertook two variants of commercial market simulations—one that analyzes the market from the perspective of U.S. firms and another that takes a more global view. The U.S.-centric simulations project the number of addressable commercial heavy lift launches that U.S. firms will win in the coming years and compare those figures with projected excess U.S. launch capacity—launch capacity remaining after NSS launches have been netted out. On the supply side, we assume SpaceX, as the current market leader, will remain in the commercial market independent of selection as an NSS provider, but that other new entrants drop if they are not selected for NSS. Both the size of the addressable commercial market and the U.S. market share are varied, with the latter dropping following the introduction of new foreign competition (the Ariane 6 and Soyuz 5 rockets). This analysis and the resulting plots help shed light on how many U.S. firms the launch market can plausibly support.

The other set of commercial market simulations centers around assessing global commercial launch capacity relative to global commercial launch demand across all possible U.S. NSS launch provider scenarios. For this analysis, aggregate commercial launch capacity includes that of Russia and Arianespace in addition to the capacity of U.S. firms. The procedure for calculating future annual Russian and Arianespace launch capacity mirrors that used in constructing U.S. firm capacity and accounts for when the Soyuz 5 and Ariane 6 rockets are expected to enter the market. Yearly addressable commercial heavy lift launch demand is drawn from RAND market forecasts. For this analysis, we take a holistic view of the launch market to identify periods of time where capacity relative to demand creates an environment that encourages either market entry by additional firms or market consolidation among existing firms. Entry and exit are both likely to affect the U.S. commercial market share and hence both have potential ramifications for how many U.S. firms the market can support in the longer term.

We turn now to in-depth discussions of the details of the NSS and commercial market simulations.

NSS Market Simulations

We begin this section by walking through the construction of one particular NSS market plot in detail. Following this examination, we discuss the various excursions from this baseline case that round out our NSS market analysis.

The plot we choose to dissect depicts yearly net NSS capacity (NSS launch capacity less NSS launch demand) from 2020 to 2030 for the case in which the Vulcan and OmegA are chosen as the two NSS launch providers under the Phase 2 contract. Our goal is to construct 1,000 randomized annual launch capacity profiles along with 1,000 randomized annual NSS launch demand profiles. We then subtract our 1,000 NSS demand profiles from our 1,000 NSS capacity profiles to obtain 1,000 yearly net capacity scenarios to determine range (and probability of occurrence within that range) of net NSS capacity across simulation iterations, by year. The random elements that vary across simulation iterations are (1) delays to the projected first launch date for each launch vehicle,[1] (2) maximum annual launch capacity for each launch vehicle, and (3) the number of NSS launches that experience a delay each year. The range of these variables is summarized in Table B.1; evidence to support the ranges is provided in the body of this report.

The first two rows in Figure B.1, delays in first launch and maximum annual capacity, relate to the capacity side of the equation. Vulcan has a projected first launch date of 2021 and is currently projected an annual launch capacity between six and 12. We begin by randomly drawing a delay in first launch, assigning equal probabilities to one-, two-, three-, and four-year delays. As an example, suppose we draw a one-year delay so that first Vulcan launch takes place in 2022. We next randomly draw a number between six and 12 (once again assigning equal probabilities to each number)

Table B.1
Range of Uncertainties Explored in the NSS Monte Carlo Analysis

Uncertainties	Range Explored	Source or Basis
NSS yearly demand slips; quantity that slips to year N+1 (with its procured launch vehicle)	[0 – 30%] of current forecast (prior slide)	RAND's analysis of accuracy of demand forecasts NSSL procurement process
U.S. firms' yearly capacity-sizing decisions	Vulcan: [6–12] yearly launches OmegA: [6–8] yearly launches New Glenn: [12–16] yearly launches Falcon Heavy: [6–20] yearly launches All SpaceX if selected: [20–24] yearly launches All SpaceX if not selected: [14–20] yearly launches	RAND discussions with Vulcan, OmegA and New Glenn; RAND analysis of SpaceX corporate announcements
U.S. new entrant first launch dates	Vulcan: 2021 + [1–4] years OmegA: 2021 + [1–4] years New Glenn: 2021 + [1–4] years Falcon Heavy: 2018	USAF provided dates for Vulcan, OmegA, New Glenn; RAND analysis of accuracy of first launch forecasts; Falcon Heavy actual first launch date
USAF contingency access to legacy launch systems	2022: Probability = 85%, Years of service regained = 3.5 2023: Probability = 85%, Years of service regained = 2.5 2024: Probability = 50%, Years of service regained = 1	RAND analysis of Phase 2 Model Contract contingency clause

SOURCE: RAND analysis as denoted in the table.

---

[1]   Unless a first launch has already occurred, as it has for such legacy vehicles as the Falcon 9 and Falcon Heavy.

to represent Vulcan annual launch capacity. Suppose we draw the number 12. We then construct an annual launch capacity profile, taking into account that first launch occurs in 2022 and maximum capacity is 12 launches per year, with the rate of buildup roughly matching that of the SpaceX Falcon 9. In this example, yearly Vulcan launch capacity from 2020 to 2030 would be: 0, 0, 2, 3, 4, 4, 5, 12, 12, 12, 12. We then replicate this process for the OmegA, whose projected first launch date is also 2021 plus a random one- to four-year delay and whose assumed annual launch capacity is between six and eight to produce a capacity profile. We then sum capacity across the two providers, by year, to obtain the NSS launch capacity. This procedure is repeated 1,000 times to represent the plausible range of NSS launch capacity should these two providers be selected for NSS Phase 2.

The capacity buildup rate for the new entrants turned out to be the dominant determinate of whether or not there would be a shortage in supply for near term NSS launches. A typical single future, representing the capacity build up for when three new entrants are selected for the Phase 2 contract, is shown in Figure B.2 for the years 2022 to 2025. The nominal demand for NSS launches in these years is four in 2022 and eight in 2023 and 2024, which would leave a shortage in both 2022 and 2023 if no delays in demand are realized.

We now turn to the demand side of our NSS market analysis. For each year with unassigned NSS launch demand in Chapter Five's Figure 5.1, we construct 1,000 random NSS demand profiles by allowing up to 30 percent of unassigned NSS pay-

Figure B.2
Example of Launch Capacity Buildup for a Three-Provider Scenario

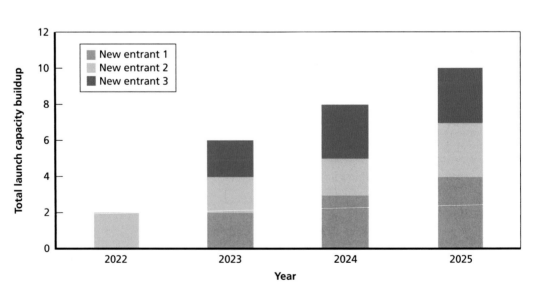

SOURCE: RAND analysis.

loads in a given year to slip to later years.[2] In a given simulation iteration, this procedure is executed for each year in our time frame. It is worth noting that launches that have slipped from one year to the next are not included in the total count when computing slip for the following year. This mimics the effect of a payload and launch vehicle slipping together and avoids double counting launch slips because of payload unavailability. This process is then repeated 1,000 times to produce our 1,000 NSS launch demand profiles in which randomness stems from the number of launches slipped each year.

Once we have 1,000 NSS launch capacity scenarios and 1,000 NSS launch demand scenarios, we subtract the latter from the former (by year) to obtain 1,000 NSS net capacity profiles. Having 1,000 different net capacity profiles means that we have 1,000 different net capacity figures for each year between 2020 and 2030 for the two-provider scenario when Vulcan and Omega are selected for Phase 2.

Our research question, however, is not how any specific Phase 2 provider selection affects the heavy lift launch market. Instead, we want to understand how a USAF decision to select two versus three providers might affect the market. Therefore, we aggregated statistics across different launch provider scenarios to produce plots of mean net capacity, along with the associated 90-percent confidence interval. We did this for all two provider cases and for all three provider cases. However, we quickly realized that there was a distinct cluster of results that did not depend on number of providers but was instead distinguished by whether only new entrants were selected for the Phase 2 contracts. Therefore, a similar process was used to aggregate statistics across those scenarios. We call these aggregated outcomes a "future set."

Figure B.3 shows the output of the aggregated results for all possible two-provider scenarios, and it is built as follows:[3]

- First, we "stack" the 1,000 net capacity profiles for each of the six possible two-provider scenarios by year.[4] This results in 6,000 net capacity values being associ-

---

[2]   To replicate our analysis, attention is required to ensure the "tail" of the probability distribution is modeled correctly. For example, suppose there are nine unassigned NSS launches in a given year. Multiplying nine by 0.3 (30%) yields 2.7; rounding up implies that a maximum of three launches could be delayed from payload unavailability. However, to model a uniform distribution of 0 to 30-percent uncertainty, the probability of three launches slipping must be 70 percent of the probability of zero, one, or two launches slipping. Solving for the probabilities that meet this constraint, yields a 27-percent chance each of zero, one, or two launches slipping and an 18-percent chance of three launches slipping.

[3]   This plot was created prior to decisions by the USAF to exercise its remaining Delta IV Heavy options under the Phase 1A contract. These options were exercised in May 2019 and August 2019 during the period of performance of this research. Therefore, the results of our final analysis—documented in the main body of this report—are slightly better than what is shown in this appendix.

[4]   Given we have no knowledge of which providers will submit bids for the Phase 2 contract, we considered all possible combination of the four possible bidders. Hence, there are six possible two-provider combinations.

Figure B.3
Net Capacity of NSS-Certified Providers in All Two-Provider Scenarios

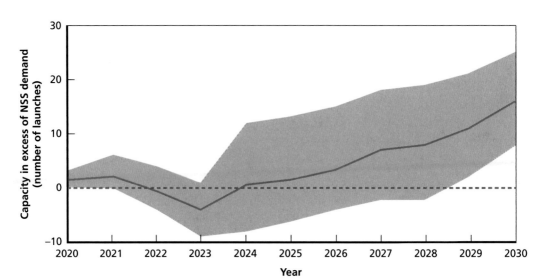

SOURCE: RAND analysis.
NOTE: Assumes nonselected providers drop or do not pursue NSS certification.

ated with each year between 2020 and 2030—1,000 for each of the six launch provider scenarios.

- Second, we compute mean net capacity within each year to obtain yearly average net capacity across all two-provider scenarios; this corresponds to the solid black line in the figure.
- Finally, we rank the 6,000 net capacity values within each year from smallest to largest and select the 5th and 95th percentile values from each year to construct our 90-percent confidence interval. These values correspond to the lower and upper ends of the gray band depicted in the figure, respectively.[5]

As mentioned in the previous subsection, we also construct histograms displaying the breakdown of all simulation outcomes for a particular set of futures for years in which NSS launch risk is high—i.e., when net capacity ranges include negative values. An example histogram of year 2023 outcomes when the USAF only has access to two new entrants is shown in Figure B.4. This histogram shows the fraction of all 1,000 simulations that resulted in a shortage of supply. The horizontal axis values range from

---

[5] The 5th percentile net capacity value in a given year is the value such that 5 percent of net capacity values in that year are lower than this number; the 95th percentile net capacity value in a given year is the value such that 95 percent of net capacity values in that year are lower than this number. Accordingly, 90 percent of simulated two-provider net capacity outcomes fall within this range—this is the 90-percent confidence interval referenced above.

−11 to −1, meaning that all 1,000 net capacity outcomes for this particular provider selection fell in this range. Taking the gray bar rising from the net capacity value −6, this plot is read as "17 percent of the 1,000 simulation (e.g., 170 of the 1,000 net capacity outcomes) had a shortage of 6 launches in 2023."[6] The heights of the gray bars sum to 1 (100 percent), meaning that the plot accounts for all 1,000 simulation outcomes.

It is the outputs of these histograms at different percentile outcomes that are used to determine the points graphed onto the risk framework used in Chapter Five to assess the impact of USAF NSS launch acquisition decisions. For example, to convert from number of launches short to the percentage of planned service years lost between 2022 and 2025 at 5 percent likelihood, we sum the fractions starting from the left-hand side until we get to 5 percent—in this case, somewhere between nine and ten launches short. Because a launch in 2023 would normally provide three years of service (2023, 2024, and 2025), we multiply this quantity by three. We then repeat this process for each year to get the total service years lost and then divide by the number of service years the USAF had planned to gain from the demand over this time period. We repeated this process for the 25 percent, 50 percent and 75 percent likelihoods to graph the line onto our risk framework (see Figure B.5).

Figure B.4
Histogram of 2023 Net NSS-Certified Capacity Simulation Outcomes

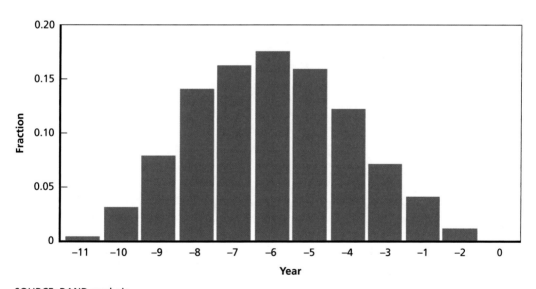

SOURCE: RAND analysis.
NOTE: Assumes nonselected providers drop or do not pursue NSS certification.

---

[6]   Although most outcomes fell in the range of four to eight launches short, this plot was generated prior to the USAF exercise of options for the remaining Delta IV Heavy legacy vehicles. Final results, documented in the main body of this report, are improved from what is shown in this appendix.

When computing the service years regained if the USAF exercises the remaining options of the Phase 1A contract and the contingency clauses of the Phase 2 contract to obtain nonpriority access to legacy systems, we added back in years of service life based on assumptions regarding the probability of successfully acquiring a legacy launch vehicle through those means and an estimate of the number of service years that could be regained through that mechanism. Note that we did not pick from a range of values in a Monte Carlo method when evaluating this particular set of futures. Therefore, we are less robust to errors in our assumptions for this portion of the analysis.

Analysis Excursions for the NSS Market

To provide the USAF with additional insight into how potential provider decisions might affect the near-term NSS market, we conducted excursions from the baseline set described in the main body of this report. the resultant shortage statistics for 2022 to 2025 at the 25th, 50th, 75th, and 95th percentile for the baseline and excursions are described in Table B.2.

Excursion A: This excursion demonstrates the sensitivity of our analysis to assumptions regarding delays in first launch dates. The run is identical to the baseline (results in Table B.2), with the exception that delays in first launch data are uniformly

Figure B.5
Percentage of Planned Service Years Lost as a Function of Likelihood for Four Different Sets of Futures

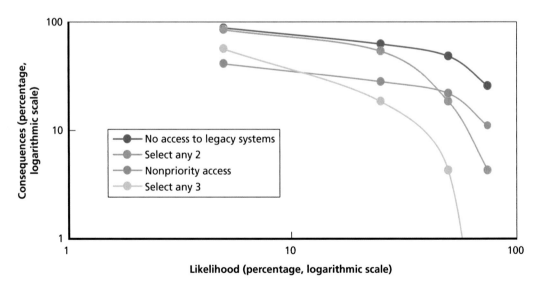

SOURCE: RAND analysis.

Table B.2
Baseline NSS Results (Single-Year Shortages)

Future Set	Year	Single-Year Shortages			
		Number of Vehicles Short			
		25%	Median	75%	95%
Legacy + any 1 or 2	2022	0	0	0	1
	2023	0	1	2	4
	2024	0	0	0	0
	2025	0	0	0	0
Any 2 or 3 new entrants, no legacy	2022	1	2	3	4
	2023	3	5	6	8
	2024	2	4	5	8
	2025	1	3	4	6
Any 2 Providers (may include legacy)	2022	0	1	3	4
	2023	1	3	5	8
	2024	0	0	4	7
	2025	0	0	3	6
Any 3 providers (may include legacy)	2022	0	0	1	3
	2023	0	1	3	6
	2024	0	0	0	4
	2025	0	0	0	2

SOURCE: RAND analysis.

distributed from zero to three years.[7] At the 50th percentile, this change in assumption decreases the number of vehicles short by one to two launches. The overall relationships between the future sets is unchanged.

Excursions B1 to B3: We also ran a set of excursions using the zero–to–three-year nominal delay in first launch date to explore how capacities might change if one of the new entrants not selected for the Phase 2 contract decided to continue to pursue NSS certification. This would allow that firm to provide contingency launch services

---

[7]   As noted earlier in this report, we found no evidence to support a range of first launch delay that included zero as an equal probability with one, two or three years of delay. We discussed including a zero-year delay at very low probability, but had no evidence from which to assign a value to the probability. After consultations within RAND, we decided to use a value range of one to four years as the baseline (having evidence for each of those values) and run this excursion to illustrate the sensitivity of our analysis to assumptions in first launch delay.

under the Phase 2 contract if selected providers were unable to meet the NSS demand. The results were relatively insensitive to *which* new entrant decided to continue; Table B.4 gives the resultant shortages from a representative excursion. Comparing the results with Table B.3 illustrates that having an additional new entrant as a contingency backup might help, but because capacity is still contingent on first launch dates and initial buildup rates, the impact is at most one launch in some years at the 50th percentile. The tails of the distribution (i.e., the 25th and 95th percentile results) are relatively unchanged. As demonstrated by these excursions, the contingency backup clause of the Phase 2 contract only provides significant risk mitigation if it allows the USAF to access a legacy provider with existing capacity.

Excursion C1: Identical to Excursion A, (i.e., using the zero–to–three-year range for launch delays) except BE-4 engine certification is slowed, delaying the Vulcan by

Table B.3
Excursion A: Single-Year Shortages if Launch Delays Are 0–3 Years

| Future Set | Year | Single-Year Shortages | | | |
| | | Number of Vehicles Short | | | |
		25%	Median	75%	95%
Legacy + any 1 or 2	2022	0	0	0	1
	2023	0	0	1	4
	2024	0	0	0	0
	2025	0	0	0	0
Any 2 or 3 new entrants, no legacy	2022	0	1	2	4
	2023	1	3	5	7
	2024	0	2	3	5
	2025	0	1	3	5
Any 2 providers (may include legacy)	2022	0	0	1	4
	2023	0	2	4	6
	2024	0	0	2	5
	2025	0	0	2	5
Any 3 providers (may include legacy)	2022	0	0	0	2
	2023	0	0	1	4
	2024	0	0	0	1
	2025	0	0	0	0

SOURCE: RAND analysis.

Table B.4
Single-Year Shortages if a New Entrant Continues to Pursue NSS Certification Independent of Selection as a Phase 2 Provider

Future Set	Year	Single-Year Shortages Number of Vehicles Short			
		25%	Median	75%	95%
Legacy + any 1 or 2	2022	0	0	0	1
	2023	0	0	0	3
	2024	0	0	0	0
	2025	0	0	0	0
Any 2 or 3 new entrants, no legacy	2022	0	1	2	4
	2023	1	2	4	7
	2024	0	1	3	5
	2025	0	1	3	5
Any 2 providers (may include legacy)	2022	0	0	1	3
	2023	0	1	3	6
	2024	0	0	1	4
	2025	0	0	1	4
Any 3 providers (may include legacy	2022	0	0	0	2
	2023	0	0	1	4
	2024	0	0	0	1
	2025	0	0	0	1

SOURCE: RAND analysis.

two years (i.e. Vulcan has an equal probability of being two or three years late) and the New Glenn by five years. Comparing the results in Table B.5 against those in Table B.3, the net effect of such a delay is that the probability of shortages is increased significantly in all years and under all scenarios.

Excursion C2: For this excursion, we have assumed that the GEM-63 engine certification is delayed, setting back the first launch of both the Vulcan and the OmegA by three years. Results are quite similar to those for the BE-4 engine delays in the years 2022 and 2023, but the later years are not as affected because both providers come online in 2023 (see Table B.6).

Table B.5
Single-Year Shortages if BE-4 Engine Development Is Delayed, Affecting Vulcan and New Glenn First Launch Dates

| Future Set | Year | Single-Year Shortages | | | |
| | | Number of Vehicles Short | | | |
		25%	Median	75%	95%
Legacy + any 1 or 2	2022	0	0	0	1
	2023	0	1	2	4
	2024	0	0	0	0
	2025	0	0	0	1
Any 2 or 3 new entrants, no legacy	2022	2	3	4	4
	2023	4	5	6	8
	2024	3	4	6	8
	2025	3	5	6	8
Any 2 providers (may include legacy)	2022	0	1	3	4
	2023	1	3	5	8
	2024	0	1	5	7
	2025	0	2	5	8
Any 3 providers (may include legacy)	2022	0	0	1	4
	2023	0	1	3	6
	2024	0	0	1	4.5
	2025	0	0	1	5

SOURCE: RAND analysis.

Commercial Market Simulations

As discussed earlier in this appendix, we analyze the commercial heavy lift launch market from two different angles: one from the perspective of U.S. firms and a second one that looks at the market more comprehensively. The demand for U.S. commercial launches is a product of the size of the addressable commercial market and the U.S. market share; therefore, we account for how the U.S. market share is expected to change as the Soyuz 5 and Ariane 6 rockets enter the market in the coming years. We then compare the demand for U.S. commercial launches with the U.S. supply in excess of NSS demand to understand local (U.S.) market dynamics. We then take a step back and examine global commercial launch capacity in relation to global commercial launch demand to shed light on the prospects of seeing additional market

Table B.6
Single-Year Shortages if GEM-63 Engine Development Is Delayed, Affecting Vulcan and OmegA First Launch Dates

Future Set	Year	Single-Year Shortages			
		Number of Vehicles Short			
		25%	Median	75%	95%
Legacy + any 1 or 2	2022	0	0	0	1
	2023	0	1	3	5
	2024	0	0	0	0
	2025	0	0	0	0
Any 2 or 3 new entrants, no legacy	2022	1	3	4	4
	2023	4	6	7	9
	2024	0	2	3	6
	2025	0	1	3	6
Any 2 providers (may include legacy)	2022	0	1	3	4
	2023	1	4	6	8
	2024	0	0	2	5
	2025	0	0	2	5
Any 3 providers (may include legacy)	2022	0	0	1	4
	2023	0	2	4	7
	2024	0	0	0	2
	2025	0	0	0	1

SOURCE: RAND analysis.

entrants during this time frame. This, naturally, involves incorporating Russian and Arianespace commercial launch capacity into aggregate capacity. In the forthcoming discussion, we take a deeper look at each of these exercises and detail the differences between them and our NSS market simulations.

Launch capacity in the simulations centered on the U.S. vantage point is constructed in nearly the same manner as in the NSS analysis described in the previous subsection. The only difference relates to SpaceX. In our commercial market analysis, we assume that if SpaceX is selected as an NSS launch provider, they have a maximum annual launch capacity between 20 and 24 and as the leading commercial provider today, they are operating at capacity (i.e., we do not apply a buildup curve). Moreover, if they are not selected as an NSS launch provider, we assume that they choose to remain in the commercial market and right size their capacity to the commercial market (14 to 20 as documented in Table B.1).[8] We add SpaceX capacity to total U.S. capacity for launch provider scenarios in which SpaceX is not chosen as one of the NSS launch providers. The other key assumptions of the commercial market analysis are shown in Table B.7.

To build our 1,000 U.S. commercial demand profiles, we must do several things.

- First, in each simulation iteration, we choose the size of the global addressable commercial heavy lift launch market by randomly picking a number between the minimum and maximum RAND-forecasted market size for each year from 2020 to 2030, assigning equal probabilities to each number within each range.
- Second, in each simulation iteration, we choose the U.S. market share in each year. We do this by randomly picking a number between the assumed range for each year.[9]
- Finally, we multiply the size of the market in each year by the U.S. market share in each year to obtain annual forecasted U.S. commercial launch demand. This constitutes one simulation iteration; we repeat this process 1,000 times to create 1,000 U.S. commercial launch demand profiles.

Table B.7
Range of Uncertainties Explored in the Commercial Market Monte Carlo Analysis

Uncertainties	Range Explored	Source or Basis
Yearly addressable demand	RAND forecast with uncertainty bounds	RAND analysis
Ariane and Soyuz yearly capacity-sizing decisions for addressable market	Ariane: [4–6] growing to [5–11] with introduction of Ariane 6 Soyuz: [4–10] growing to [8–14] with introduction of Soyuz 5	Assumes providers' size to their historical range of addressable launches per year
Ariane and Soyuz new entrant first launch dates	Ariane 6: 2020 + [0–3] years Soyuz 5: 2022 + [1–4] years	Provider's announced dates and RAND's analysis of accuracy of first launch forecasts
U.S. market share	[30%–67%] before Ariane 6 and Soyuz 5 come to market [30%–50%] after either come to market [20%–30%] after both come to market	Assumes market dynamics return to the equilibrium state of the early to mid-2000s

SOURCE: RAND analysis as indicated in the table.

---

[8]   The reason we do not impose a buildup process on SpaceX on the commercial side is due to the current commercial viability of the Falcon 9.

[9]   The U.S. market share range is varied based on when Arianespace and Russia bring new vehicles to market as described in the main body of this report.

Upon constructing the 1,000 launch capacity and demand profiles, we compute total U.S. excess launch capacity by subtracting both NSS launch demand (described in the previous subsection) and U.S. commercial launch demand from U.S. launch capacity. We do this separately for each NSS launch provider scenario; the plots take the same form as those described in the previous sub-section. Furthermore, Figure B.6 directly plots the mean and 90-percent confidence interval of U.S. commercial launch demand over time, taking into account the expected change in U.S. market share following the entry of the Soyuz 5 and Ariane 6 rockets. Mean U.S. commercial launch demand (the solid black line in the figure) is computed by calculating the within-year average across the 1,000 simulations. The 90 percent confidence interval (the gray shading in the figure) is constructed by plotting the 5th and 95th percentile U.S. commercial launch demand values in each year over time.

The key difference between the U.S.-focused and globally oriented commercial market analyses lies in how total launch capacity is computed. Total U.S. launch capacity consists of the capacity of U.S. launch providers; total global commercial launch capacity further includes Russian and Arianespace capacity for the purposes of this study. Both have a significant commercial market share and we expect that share to increase as both introduce new vehicles better suited to the demands of today's market (the Soyuz 5 and Ariane 6). Our approach assumes the initial capacity for each is in the range of their recent market share until the new vehicle is brought online. At that point we assume they will increase their capacity, with the expectation of regaining market share approximate

Figure B.6
Forecasted Addressable Commercial Launches Won by U.S. Firms

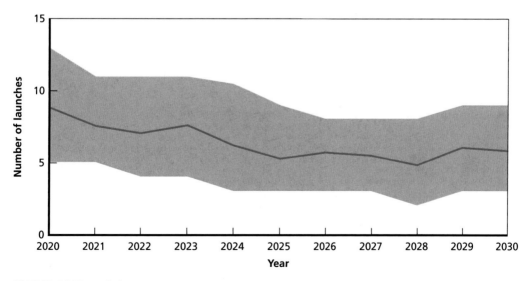

SOURCE: RAND analysis.

to their historical market share. As with new U.S. entrants, we compute the year of first launch from their predicted first launch date plus randomized uncertainty.

Thus, for a given simulation iteration, we construct Russian commercial capacity as follows. First, we randomly select a Soyuz 5 launch delay between one and four years. We then select a random number between four and ten to represent their annual launch capacity prior to actual Soyuz 5 first launch and randomly select a number between eight and 14 to represent annual launch capacity from that date onward. We repeat this process 1,000 times to acquire 1,000 annual Russian commercial launch capacity profiles. Arianespace commercial launch capacity is constructed in the same fashion.

Global commercial launch capacity is then constructed as the sum of Russian, Arianespace and total U.S. capacity, minus U.S. NSS launch demand. U.S. capacity varies depending on each two and three NSS launch provider scenario. We use global commercial capacity in conjunction with global commercial demand to produce an assortment of graphical outputs. Our primary outputs consist of plots depicting the yearly minimum-to-maximum global commercial capacity ranges across 1,000 simulations. As with our NSS market analysis, we also average several different future sets (e.g., all scenarios in which two providers are chosen or all scenarios in which three providers are chosen) and plot mean global commercial capacity, along with the associated 90-percent confidence intervals as shown in Figure 6.2 in Chapter Six. On these plots, we look at two thresholds—the first is when supply is less than 1.5 times market demand, the value at which buyers told us they would seek out a new supplier. Therefore, this threshold is an indicator that additional foreign competitors might enter the market. The second threshold is when supply exceeds 2.5 times the market demand (less NSS demand), indicating commercial market saturation. Although state supported providers may stay in the market despite this over supply, it is less certain that U.S. firms could remain profitable at these levels. Therefore, we use this threshold as an indicator of possible U.S. supplier consolidation.

Analyses Conducted for the Commercial Market Assessment

We conducted a variety of commercial market analyses. In each, 1,000 simulations were undertaken for each year between 2020 and 2030 (inclusive). A summary of each analysis is listed below. Following brief descriptions of each, we report simulation results on commercial market saturation as a function of the year in which global commercial supply exceeds 2.5 times global commercial demand. We conclude the section by including Stata code, which may be used to replicate our analyses.

- Projected U.S. total net launch capacity (i.e., launch supply minus U.S. NSS and commercial launch demand) over time, under each contract award scenario, assuming the U.S. share of addressable commercial heavy lift launches remains within its recent historical range.

- Projected U.S. total net launch capacity (i.e., launch supply minus U.S. NSS and commercial launch demand) over time, under each contract award scenario, assuming the U.S. share of addressable commercial heavy lift launches changes as the Ariane 6 and Irtysh/Soyuz 5 vehicles enter the market.
  - Additionally, we examined the mean (5th percentile and 95th percentile of U.S. addressable commercial heavy lift launch demand over time) assuming the U.S. share of addressable commercial heavy lift launches changes as the Ariane 6 and Irtysh/Soyuz 5 vehicles enter the market.
- Projected global net commercial capacity over time, under each contract award scenario, as well as projected global commercial launch supply and demand— separately—over time, under each contract award scenario.
  - Additionally, we examined the mean, 5th percentile and 95th percentile of global commercial launch supply over time, where the mean and percentiles were calculated across all two-provider scenarios and, separately, all three-provider scenarios.

Table B.8 displays, for each possible combination of contract awardees, the first year in which median forecasted global commercial heavy lift launch supply meets or exceeds 250 percent of addressable commercial demand. The median is computed, by year, across 1,000 simulation iterations for each contract award scenario. These results demonstrate that market saturation is largely independent of near-term USAF decisions regarding the Phase 2 contract, but they might occur quite soon (as early as 2022) under some scenarios and circumstances.

Table B.8
Market Saturation as a Function of USAF Decisions on Phase 2 Providers

Contract Recipients	First Year in Which Ratio of Global Supply to Demand ≥ 2.5 at Each Percentile		
	25th	50th	75th
ULA and Blue Origin	N/A	N/A	2027
ULA and NG	N/A	N/A	2028
ULA and SpaceX	N/A	N/A	2028
Blue Origin and NG	N/A	N/A	2027
Blue Origin and SpaceX	N/A	N/A	2028
NG and SpaceX	N/A	N/A	2028
ULA, Blue Origin, and NG	N/A	2028	2027
ULA, Blue Origin, and SpaceX	N/A	2028	2022
ULA, NG, and SpaceX	N/A	N/A	2022
Blue Origin, NG, and SpaceX	N/A	N/A	2025

SOURCE: RAND analysis. Supply is computed as the sum of contract recipients' supply, Arianespace's supply, Russia's supply and SpaceX's supply (if SpaceX was not a contract recipient) minus U.S. NSS demand. Addressable commercial heavy lift launch demand is drawn from RAND forecasts. The 25th percentile, 50th, and 75th percentile values are computed, for each year, across 1,000 simulation iterations.

Learning from History

This appendix contains summaries regarding the history of the global heavy lift launch market and the providers who have historically competed in it. We provide these summaries in the hopes of offering perspective on the dynamics of both the NSS and addressable commercial markets and their interactions.

Foreign Government Investments in Heavy Lift Launch

Japan

Japan's government funds improvements by Mitsubishi Heavy Industries to the Japanese Aerospace Exploration Agency (JAXA) H-II and H-III series of heavy lift launch vehicles at a moderate pace and gives preference to these domestic vehicles when launching domestically manufactured or funded satellites and for JAXA deliveries to the ISS.[1] Therefore, Japan's investments are directly comparable with the U.S. investments shown in Figure 2.3 in Chapter Two.

Europe and India

European and Indian investments are less directly comparable because they include investments in small to medium lift launch vehicles that are not the subject of this report. The European investments include direct European Space Agency funding of Ariane 6 and Vega development programs at Arianespace, a promised quantity of non-addressable launches for those launchers, and the ability for the agency to run deficits on the order of €200 million per year, covered by the European Commission.[2] The

---

[1] JAXA, *About H-IIB Launch Vehicle, and H-IIA Launch Vehicle*, 2019; Mitsubishi Heavy Industries, *H-IIA User's Manual*, Version 4.0, Tokyo, Japan: Mitsubishi, February 2015.

[2] Dominique Gallois, "Ariane 6, un Chantier Européen pour Rester dans la Course Spatiale," *Le Monde*, December 2, 2014; European Space Agency, *ESA Annual Report*, The Netherlands, AR-2016, December 2016; Thierry Dubois, "Arianespace Needs Both Commercial And Institutional Orders," *Aviation Week and Space Technology*, January 16, 2019.

Vega is a small to medium lift launch vehicle, but we are unable to separately estimate and remove this investment.

India's government has budgeted for the Indian Space Research Organization (ISRO) to build 30 PSLV and ten GSLV Mk III (also known as LVM3) launch vehicles, many to be resold as launch services on the commercial market.[3] The PSLV is a small to medium lift launch vehicle, but again we are unable to separately estimate and remove this investment.

China

Our estimate of China's investments in launch vehicle development are derived from the budget of the Chinese National Space Agency (CNSA).[4] CNSA is developing a complete series of new Long March rockets, including heavy lift launch vehicles and reusable boosters (often referred to as the Long March 5/7 and Long March 8/9 families of vehicles), as well as improvements to the Long March 3 family.[5] Much of this investment is undoubtedly driven by China's space exploration goals, including moon landings and the building of its own domestic space station. Only the Long March 3 family is used to compete in the global commercial heavy lift launch market.

Russia

Russia has invested in the improved Soyuz 2 in cooperation with Arianespace. Russia is also currently investing in the Soyuz 5, also called Irtysh.[6] Roscosmos' newest launch vehicle series (the Angara A1 and A5) has been under development since the 1990s to replace the Proton, Rokot, Kosmos, Cyclone, Dnepr and Zenith launch vehicles. Only the Proton replacements are heavy lift launch vehicles, but we are unable to separate out the investments by class of vehicle.[7]

---

[3]   "India Approves US$1.574 Billion Funding for ISRO's 30 PSLV & 10 GSLV Flights," *Spacetech*, June 6, 2018; Space Launch Report, "LVM3 (GSLV Mk 3)," webpage, last updated July 22, 2019a; Government of India, "ISRO Budget at a Glance," webpage, undated.

[4]   Becki Yukman, "International Space Budget Data," in *The Space Report*, Colorado Springs, Colo.: Space Foundation, 2019.

[5]   Bradley Perrett, "Chinese Working On Giant Engine For Long March 9," *Aviation Week and Space Technology*, March 12, 2018a; Bradley Perrett, "Long March 8 Reconfigured For Reuse," *Aviation Week and Space Technology*, October 26, 2018b; Bradley Perrett, "Second Version Of Long March 7 Awaits Approval," *Aerospace Daily and Defense Report*, 2018c.

[6]   "Russia to Allocate $1.5Bln to Federal Space Program in 2017—Draft Budget Plan," *Sputnik News*, October 29, 2016.

[7]   Khrunichev State Research and Production Space Center, "Angara Launch Vehicles Family," webpage, undated; Irene Klotz, "Game On," *Aviation Week and Space Technology*, April 9, 2018, pp. 44–47.

Future Foreign Government Investments

Foreign governments have and are likely to continue to make significant investments in domestic launch capabilities to meet national security requirements and support other national goals related to space. Foreign governments are also likely to provide continued financial support for launch providers to compete more effectively in the global commercial market to strengthen their domestic capabilities. Moreover, the addressable segment of the global commercial market is likely to remain, or perhaps become more, constrained because of national policies from both the U.S. and foreign governments to ensure that domestic launch providers can effectively compete for at least a subset of commercial launches, but which also create limits on the global market share launch service providers can acquire.

The Formation of ULA

In May 2005, Lockheed Martin's Atlas and Boeing's Delta launch vehicles were the only two providers of NSS launch. In an effort to reduce the cost of providing launch services to the USAF and NASA, these firms petitioned the Federal Trade Commission (FTC) for the merger of their respective Atlas and Delta launch vehicle programs into the ULA. The two companies justified their petition by demonstrating that their respective launch programs were operating in the red and, without the merger, one would be forced to exit the launch market. The USAF would then be in violation of U.S. Code 10, Section 2273, which requires "the availability of at least two space launch vehicles (or families of space launch vehicles) capable of delivering into space any [. . .] national security payload." Moreover, Lockheed Martin and Boeing argued that the merger would increase mission assurance by combining the expertise of the two companies. Although the U.S. Department of Defense[8] and Northrop Grumman registered official protests, the FTC approved the merger on October 3, 2006,[9] and ULA was formed on December 1, 2006.

Although the cost of providing launch services was high in the years leading up to the formation of the ULA, variable demand for launch services by Lockheed Martin and Boeing, shown in Figure C.1, may have also exacerbated the problems these companies were facing. From 2001 to 2006, the total demand for both companies fluctuated significantly from 19 launches in 2003 to eight launches in 2005. Moreover, annual demand was not always evenly divided between the two companies;

---

[8]   Kenneth Krieg, the Defense Undersecretary for Acquisition, Technology and Logistics at the time, stated in a memorandum that ULA "will almost certainly have an adverse effect on competition, including higher prices over the long term as well as diminution in innovation and responsiveness" (see "ULA Deal Cleared, but Not Final," *SpaceNews*, June 29, 2004).

[9]   *Aviation Week and Space Technology*, "The U.S. Federal Trade Commission Has Approved the Merger of Lockheed Martin's and Boeing's Government Launch Business," October 9, 2006.

Figure C.1
Lockheed Martin and Boeing Launch Manifests, 2001–2006

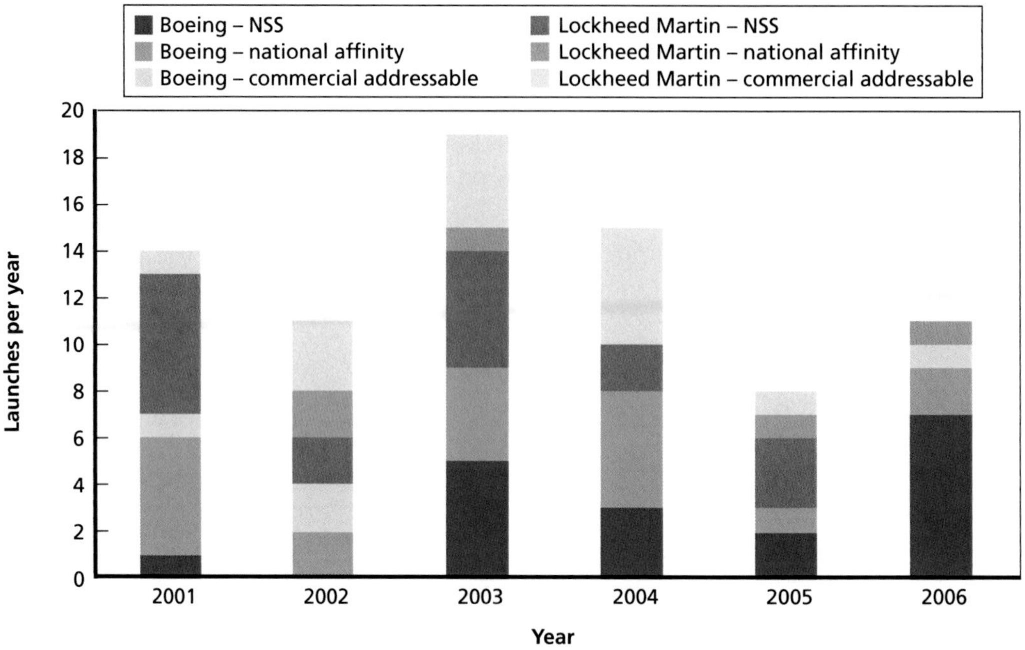

SOURCES: McCartney et al., 2006; FAA, 2013; FAA, 2018.

for example, Boeing had only three of eight launches in 2005 and Lockheed had only one of eleven launches in 2006. The combination of high cost and large fluctuations in demand would have made it difficult for Lockheed Martin and Boeing to maintain stable business practices and compete effectively in the market.

Upon the formation of ULA, Lockheed Martin and Boeing consolidated much of their infrastructure to reduce overhead costs and retained Lockheed's facilities in Colorado for administrative and engineering activities, Boeing's factory in Alabama for major assembly and integration operations, and both companies' launch facilities at Cape Canaveral Air Force Station in Florida and Vandenberg Air Force Base in California. Consolidation of the companies' infrastructures likely reduced the combined launch capacity after the merger (see Figure C.2). From 2001 to 2006, Lockheed Martin and Boeing had a combined average of approximately 13 launches a year; in the years since, ULA has had an average of approximately 11 launches a year.

Moreover, the kinds of launches (summarized in Table C.1) have also shifted since the merger, and ULA has had an average of approximately 0.5 more NSS launches per year and approximately one fewer national affinity and 1.5 fewer commercial launches per year. Thus, ULA became more of a pure Strategy C provider, as defined in Chapter Four of this report. Although Lockheed Martin competed in the heavy lift launch market using Strategy B and Boeing using Strategy C, ULA has a higher percentage

Figure C.2
Lockheed Martin and Boeing/ULA Launch Manifests, 2001–2018

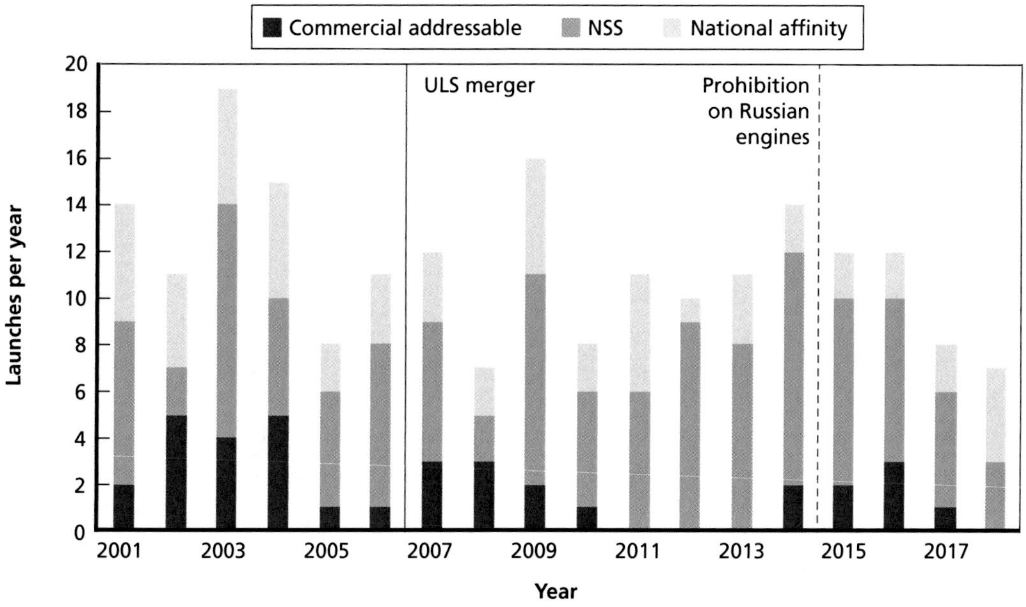

SOURCES: McCartney et al., 2006; FAA, 2013; FAA, 2018.

Table C.1
Lockheed Martin, Boeing, and ULA Launch Manifests, 2001–2018

Launch Market	LM Average Annual Launches	Boeing Average Annual Launches	ULA Average Annual Launches	Change in Average Annual Launches
NSS	3.0	3.0	6.5	+0.5
National affinity	0.8	3.2	2.8	−1.2
Commercial addressable	2.3	0.7	1.4	−1.6
Total	6.2	6.8	10.7	−2.3

SOURCES: McCartney et al., 2006; FAA, 2013; FAA, 2018.

of NSS launches than either of the prior two companies (60 percent compared with Lockheed Martin with about 50 percent and Boeing with about 45 percent) and thus has a greater reliance on the U.S. government to sustain its core business.

The formation of ULA did lead to higher prices for NSS launch services. However, prior to the merger, Lockheed Martin and Boeing NSS launch endeavors were operating at a loss and the prices for launch services did not reflect actual costs. Given both companies' small share of the addressable market, the U.S. government was the predominant buyer. This monopsony, and the U.S. government's sensitivity to price, may have led to lower-than-sustainable prices for launches services as the two providers tried to outbid each other for launch contracts. The formation of ULA balanced a single buyer versus a single provider, potentially changing the power dynamics within the U.S. segment of the launch market. We assume today's higher prices are more reflective of the actual costs of launch services. To avoid another case like ULA's, the U.S. government must remain vigilant in ensuring launch service providers over whom they have monopsony power do not perpetually sell at a loss.

The Development of the Antares

In December 2008, NASA awarded contracts to Orbital Science, for the Antares launch vehicle (previously known as Taurus II), and SpaceX to provide commercial resupply services (CRS) to the ISS. At the time of contract award, Orbital had not yet ground tested the Antares engines for either the first or second stage. In December 2009, 12 months after contract award, Orbital successfully ground tested the second stage engine, the Castor 30; three months later, in March 2010, the first-stage engine, the AJ26/NK-33, was also successfully ground tested. It was then 35 months before Orbital was able to perform a hot fire test of the entire first stage.

First test was successfully completed in February 2013, 50 months after contract award, and first flights occurred shortly thereafter. In April 2013, Orbital completed the first test flight of the Antares with a Cygnus Payload Simulator. Five months later, in September 2013, a second test flight was launched with a Cygnus Commercial Orbital Transportation Services (COTS) Demonstrator and successfully rendezvoused and docked with the ISS. On January 9, 2014, 61 months after contract award and 28 months after the original first launch date, Orbital successfully launched the first CRS mission to the ISS.

During the launch of the third CRS mission, on October 28, 2014, the Antares' first-stage engine, the AJ26, failed shortly after liftoff and resulted in a launch failure. A second AJ26 had also failed during a test fire five months earlier. This string of engine failures led Orbital to replace the first-stage engine with the RD-181, another manufactured Russian engine. Orbital received the first batch of RD-181 engines in July 2015 and completed a successful ground test in May 2016. On October 17, 2016, Orbital successfully launched the fourth CRS mission using the new RD-181 first-stage engine. Since 2016, Orbital has successfully launched four additional CRS missions to the ISS.

The history of the Antares provides some insight into launch vehicle development. First, the longest development time occurred between the engine ground tests and the hot fire test of the first stage, which was approximately three years (see Table C.2). Within a year of the first-stage test, the Antares completed two flight tests and its first

Table C.2
Antares Launch Vehicle History

Year	Launches	Notes
2008	0	• NASA CRS contracts awarded in December
2009	0	• Castor 30 second stage engine successfully ground tested in December
2010	0	• AJ36 first-stage engine successfully ground tested in March • Original first launch date in December
2011	0	
2012	0	
2013	2	• Successful first-stage hot fire test in February • Flight tests in April and September
2014	3	• First CRS mission launch in January • Launch failure due to the AJ26 first-stage engine in October
2015	0	• First batch of new RD-181 first-stage engines received
2016	1	• New RD-181 first-stage engine successfully ground tested in May • First launch with the new RD-181 first-stage engine in October
2017	1	
2018	2	
2019	1	• Launches as of June 2019

SOURCE: RAND analysis.
NOTE: Blank cells = Nothing significant to note that year.

CRS mission to the ISS. Second, the Antares experienced a delay in first launch of 28 months, which is close to the 31-month average delay reported by NASA.[10] Thus, predicted first launch dates do not appear accurate until near the time of the first-stage hot fire test or until the prediction is made for within the next six to nine months. However, even near-term predictions appear to be optimistic by three to six months.

The Sea Launch Story

Sea Launch formed in 1995 as a consortium of U.S., Norwegian, Ukrainian, and Russian companies, with Boeing having a leading stake of 40 percent.[11] Sea Launch completed its first launch in 1999 and, through 2014, executed a total of 32 success-

---

[10] GAO, 2010.

[11] Paula Korn and Julie Fornaro, "One World, One Platform: For Sea Launch's Multinational Crew, a Cruise on the Pacific Ocean Means Getting Ready for a Rocket's Liftoff," *Boeing Frontiers*, Vol. 1, No. 2, 2002.

ful launches, in addition to one partial success and three failed launches.[12] During its operation, Sea Launch's primary customers were commercial communication service providers that used its three-stage launch vehicle, the Zenit-3SL, to launch heavy communication satellites to GEO; thus, operating as a nearly pure Strategy A provider as demonstrated in Figure C.3.[13]

The challenges Sea Launched faced and its ultimate exit from the launch market stemmed from a string of launch failures and ensuing financial troubles, increasing competition in the launch market, and growing geopolitical tensions. Early success for Sea Launch resulted in the launch service provider winning multiple contract awards and capturing approximately 30 percent of the global market share in 2001.[14] However, in 2007, a rocket exploded on the pad during launch and Sea Launch halted all operations, which resulted in the loss of multiple launch contracts.[15] Two years later, on June 22, 2009, Sea Launch filed for Chapter 11 bankruptcy protection and the con-

Figure C.3
Sea Launch Used Strategy A, Specializing in the Addressable Market

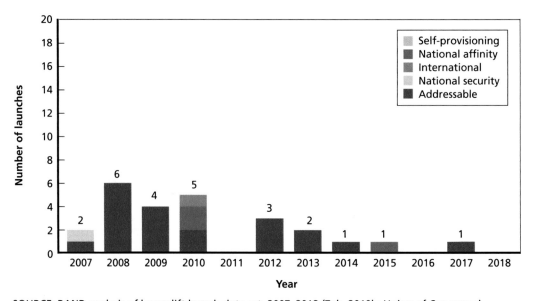

SOURCE: RAND analysis of heavy lift launch data set, 2007–2018 (Zak, 2019b; Union of Concerned Scientists, 2019; U.S. Department of Transportation, Bureau of Transportation Statistics, 2017; World Bank, undated; World Bank, 2018).

---

[12] "What Happened to Sea Launch," *Space Daily*, September 7, 2016.

[13] "What Happened to Sea Launch," 2016.

[14] Korn and Fornaro, 2002.

[15] FAA, *Semi-Annual Launch Report: Second Half of 2009*, Washington, D.C., 2009.

sortium announced it would return to normal operations after a reorganization.[16] The following year, RKK Energia, a Russian company, increased its holdings in Sea Launch from 25 percent to 95 percent to become the majority stakeholder while Boeing and other U.S. companies remained as minority stakeholders.[17]

After filing for bankruptcy and the change in majority stakeholder, Sea Launch was able to regain 11 percent and 18 percent of the addressable commercial market in 2011 and 2012, respectively.[18],[19] However, Sea Launch experienced another launch failure in 2013.[20] Later in the year, Boeing left the consortium and sued Energia and its two Ukrainian partners claiming the companies would not repay $355 million worth of investments Boeing made in Sea Launch.[21]

In addition to the launch failures and financial issues, Sea Launch experienced further challenges because of increasing competition from SpaceX and other launch service providers, as well as "the lack of any government customer ensuring consistent demand for Zenit rockets."[22],[23] In 2014, Russia annexed Crimea and the geopolitical conflict in the region increased tensions between Sea Launch's Russian and Ukrainian partners.[24] Sea Launch decided to halt operations in 2014 and reduced its staff, given the company had no scheduled launches until 2015.[25] Ultimately, Sea Launch's future launches never materialized and its last launch occurred in May 2014.[26] After halting operations, Energia also sought a buyer for Sea Launch; the S7 Group, a Russian aviation conglomerate, agreed to purchase Sea Launch and the deal was finalized in 2018.[27],[28] The S7 Group reportedly plans to resume launch operations.[29]

---

[16] "What Happened to Sea Launch," 2016.

[17] Anatoly Zak, "Sea Launch Venture," *RussianSpaceWeb.com*, February 4, 2007.

[18] FAA, *Commercial Space Transportation: 2011 Year in Review*, Washington, D.C., January 2012.

[19] FAA, *The Annual Compendium of Commercial Space Transportation: 2012*, Washington, D.C., February 2013.

[20] Peter B. de Selding, "Sea Launch Rocket Failure Destroys Intelsat IS-27 Satellite," *SpaceNews*, February 1, 2013.

[21] Zak, 2007.

[22] Caleb Henry, "Sea Launch CEO Sergey Gugkaev to Leave Company when S7 Purchase Closes," *SpaceNews*, March 13, 2018a.

[23] Jim Sharkey, "Sea Launch Reduces Staff Due to Lull in Launches," *SpaceFlight Insider*, August 27, 2014.

[24] Henry, 2018a.

[25] "What Happened to Sea Launch," 2016.

[26] "What Happened to Sea Launch," 2016.

[27] S7 Space, "Sea Launch," webpage, undated.

[28] S7 Space, undated.

[29] Caleb Henry, "S7 Closes Sea Launch Purchase, Future Rocket TBD," *SpaceNews*, April 17, 2018b.

The Proton Story

The Proton, or UR-500, began conceptually as an ICBM in the 1960s and eventually became Russia's "heaviest-lifting orbital launch vehicle," completing its first launch in 1965.[30],[31] Throughout the 1960s and 1970s, the Proton-K supported numerous Soviet space missions and, in the 1970s, successfully executed 46 out of 59 launches.[32] Similar to other Russian launch vehicles, the Proton typically launches from the Baikonur Cosmodrome, a launch facility in Kazakhstan.[33] Improvements in launch reliability eventually led to the formation of the International Launch Services (ILS) in 1995, a consortium of NPO Energia, Khrunichev Enterprises, and Lockheed Martin, to bring the Proton-K and -M launch vehicles to the commercial market; ILS later became a subsidiary of Khrunichev Space Center.[34]

The newer Proton-M launch vehicle completed its first launch in 2001 and has completely replaced the Proton-K. Over the years, improvements to the Proton-M have been made, including "weight reduction, better propellant management, engine thrust increases, and heavier propellant loads."[35] The Proton-M had a perfect record of ten launches from 2001 to 2005 before it experienced its first failure, in contrast to the more accident-prone Proton-K. However, from 2006 to 2015, the Proton-M experienced numerous failures with at least one per year (except in 2009) because of various production errors and technical failures.[36],[37] Despite this string of launch failures, the Proton remained the global market leader for commercial heavy lift launch from 2008 to 2012 and continued to operate as a Strategy B provider as seen in Figure C.4. However, since 2012, the Proton has steadily lost market share for commercial launches to SpaceX's Falcon 9.[38] The Proton Medium, a two-stage launch vehicle that is a variation of the Proton-M, was designed to be competitive with the Falcon but has thus far failed to recapture market share.[39]

---

[30]   Space Launch Report, "Proton Data Sheet," webpage, last updated August 5, 2019b.

[31]   Space Launch Report, 2019b.

[32]   Space Launch Report, 2019b.

[33]   FAA, 2018.

[34]   ILS, "Creation of International Launch Services," webpage, June 10, 1995; ILS, "About Us," webpage, undated.

[35]   FAA, 2018.

[36]   In 2016, there were difficulties during a launch that led to the grounding of the Proton-M for a year.

[37]   FAA, 2018.

[38]   RAND analysis of heavy lift launch data set (2007–2018).

[39]   FAA, 2018.

Figure C.4
Proton Uses Strategy A/B, Specializing in the Addressable Market with Some NSS Launches

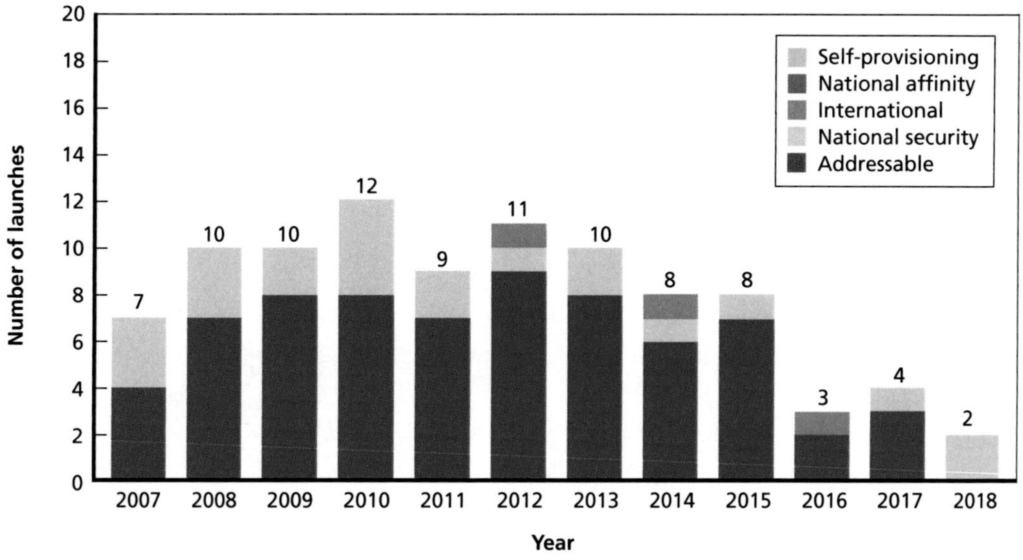

SOURCE: RAND analysis of heavy lift launch data set, 2007–2018 (Zak, 2019b; Union of Concerned Scientists, 2019; U.S. Department of Transportation, Bureau of Transportation Statistics, 2017; World Bank, undated; World Bank, 2018).

These developments reflected greater struggles for the Russian space sector in the 2010s. Brain drain of Russian space expertise was also a significant concern.[40] Thus, the United Rocket and Space Corporation was created in 2013 and aimed to "renationalize the space sector."[41] Russia also gradually doubled its spending on space to more than $5 billion in 2013.[42] In 2015, the United Rocket and Space Corporation combined with the Federal Space Agency Roscosmos to form Roscosmos Space Corporation. The Deputy Prime Minister of Russia stated, "The second stage of reform has begun. Not only the industry [as it was in the case of the United Rocket and Space Corporation], but the whole space sector is being reformed."[43]

ILS has also taken steps to compete more effectively in the global launch market and is developing the Angara family of launch vehicles, which includes the Angara 1.2, Angara 3, and Angara 5. The Angara 5 is meant to be the replacement for the Proton-

---

[40] Alissa de Carbonnel, "Russia Bets on Sweeping Reform to Revive Ailing Space Industry," Reuters, December 26, 2013.

[41] Tomasz Nowakowski, "Russia Dissolves Its Federal Space Agency, What Now?" *Space Flight Insider*, December 30, 2015.

[42] de Carbonnel, 2013.

[43] Caleb Henry, "Russia Merges United Rocket and Space Corporation with Roscosmos," *SatelliteToday*, January 23, 2015.

M and successfully completed its first test fight in 2014. The Angara 5 is expected to hold up to 7,500 kg, be capable of launching two small satellites into GEO (a current capability of the Proton-M), and launch from the Plestesk Cosmodrome approximately 500 miles north of Moscow.[44] The Angara 5 has a planned first launch date of 2020 and launches are estimated to cost $100 million.[45,46] In 2019, ILS announced it would exist under Glavkosmos, a subsidiary of the Roscosmos State Corporation, which GK Launch Services (the provider of the Soyuz 2 launch vehicle) also exists under.[47]

Comparing the histories and launch failures of Proton and Sea Launch in Figures C.5 and C.6, Proton experienced a string of failures starting in 2006 but did not begin to lose market share until eight years later. In contrast, Sea Launch experienced a launch failure in 2007 and two years later was forced to file for bankruptcy; after a second launch failure in 2013, Sea Launch was forced to exit the launch market entirely. One of the most significant differences between Proton and Sea Launch includes the presence of governmental support. Proton has and continues to receive substantial support from the Russian government, which has enabled Proton to survive its string of launch failures. Proton launched only twice in 2018 and both were for national security, which allowed Proton to continue in the market under Strategy C. The Russian government highly values and prioritizes domestic launch capabilities. Thus, Russian launch service providers are likely to continue to compete in the market and survive even the harshest economic and political environments.

The Durability of Soyuz[48]

The Russian Soyuz, meaning "Union," is the most frequently used launch vehicle in the world and is one of the most reliable launch vehicles on the market. Since its creation, the Soyuz launch vehicle has had an estimated 1,700 to 1,900 launches.[49] Over approximately the same period, the U.S. equivalent, the Atlas, has had 603 launches.[50]

---

[44] FAA, 2018.

[45] FAA, 2018.

[46] FAA, 2018.

[47] ILS, "ILS Opens New Era with New Launch Pricing; Will Operate Under Glavkosmos," webpage, April 12, 2019a.

[48] Note that a piloted spacecraft exists of the same name. The Russian Soyuz spacecraft is "the longest operational manned spacecraft programme in the history of space exploration" per European Space Agency, "Launch Vehicles: Soyuz MS Spacecraft," webpage, December 20, 2018. This section focuses solely on the Soyuz launch vehicle.

[49] Arianespace, "Technical Overview: Soyuz," webpage, undated-f.

[50] Christian Lardier and Stefan Barensky, *The Soyuz Launch Vehicle: The Two Lives of an Engineering Triumph*, New York: Springer Science+Business Media, 2013, p. xi.

Figure C.5
Timeline of Proton Successes and Failures

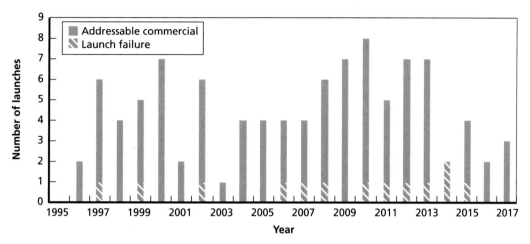

SOURCE: RAND analysis of heavy lift launch data set, 2007–2018 (Zak, 2019b; Union of Concerned Scientists, 2019; U.S. Department of Transportation, Bureau of Transportation Statistics, 2017; World Bank, undated; World Bank, 2018).

Figure C.6
Timeline of Sea Launch Successes and Failures

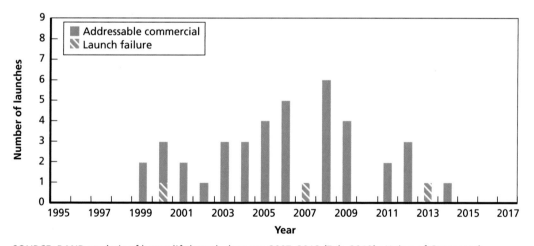

SOURCE: RAND analysis of heavy lift launch data set, 2007–2018 (Zak, 2019b; Union of Concerned Scientists, 2019; U.S. Department of Transportation, Bureau of Transportation Statistics, 2017; World Bank, undated; World Bank, 2018).

As a result of its few failures over the past six decades, the Soyuz is known as a "reliable, efficient, and cost effective solution for a full range of missions."[51]

The Soyuz is derived from the first ICBM (the R-7) and the Vostok launch vehicle, which took the first satellite (Sputnik) and first Soviet men into space.[52,53,54] The Soyuz was first launched on November 28, 1966, and the first manned launch occurred on April 23, 1967.[55] Throughout the 1970s and 1980s, the manufacturer of Soyuz launch vehicles was the Samara Space Center, later known as the TsSKB-Progress.[56] Since then, various instantiations of the Soyuz launch vehicle have been developed, including the Soyuz-U (launched in 1973), Soyuz-U2 (1982), Soyuz-FG (2001), Soyuz-2.1a (2004), and Soyuz-2.1b (2006).[57] The various instantiations of the Soyuz rocket have had different missions over the decades; the Soyuz-2.1a has transported cargo, the Soyuz-FG has transported astronauts, and the Soyuz-FG/Gregat has been used for commercial operations through a joint European and Russian venture.[58] The Soyuz design depends on the variant; however, most are three-stage launch vehicles with "RP-1 (a kerosene-based fuel) and liquid oxygen as the propellant."[59]

On October 21, 2011, a Soyuz rocket launched for the first time from a spaceport other than the Russian launch facilities at Baikonur and Plesetsk.[60] The Soyuz-ST, a variant of the four stage Soyuz-2, was launched by Arianespace from the Guiana Space Center in French Guiana and carried the first two satellites of the European global navigation satellite system (Galileo).[61,62] In 2011, the international use of the Soyuz further expanded as the United States ended the space shuttle program, which led to the Soyuz becoming the primary launch vehicle to transport astronauts and cargo to the ISS.[63] For the past decade, the Soyuz has operated primarily in nonaddressable markets using Strategy C as seen in Figure C.7.

---

[51]   Elizabeth Howell, "Soyuz Rocket: Russia's Venerable Booster," *Space.com*, April 12, 2018.

[52]   Lardier and Barensky, 2013, p. xi.

[53]   Lardier and Barensky, 2013, p. 3.

[54]   Howell, 2018.

[55]   European Space Agency, "Soyuz Launcher," *Human Spaceflight* blog, May 2005.

[56]   Space Launch Report, "R-7/Soyuz Data Sheet," webpage, last updated September 26, 2019c.

[57]   Anatoly Zak, "The Soyuz Rocket Family," *RussianSpaceWeb.com*, October 22, 2019c.

[58]   Howell, 2018.

[59]   Howell, 2018.

[60]   Arianespace, undated-f.

[61]   European Space Agency, "Soyuz," webpage, undated.

[62]   European Space Agency, undated.

[63]   Howell, 2018.

Figure C.7
Soyuz Uses Strategy C, Specializing in ISS Launches

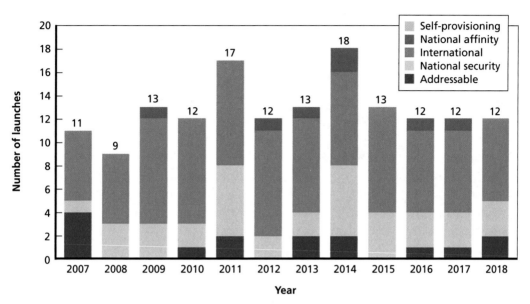

SOURCE: RAND analysis of heavy lift launch data set, 2007–2018 (Zak, 2019b; Union of Concerned Scientists, 2019; U.S. Department of Transportation, Bureau of Transportation Statistics, 2017; World Bank, undated; World Bank, 2018).

Of the launch failures that have occurred in the past 20 years, only a few have been attributed to the Soyuz rocket.[64] Five Soyuz launch failures occurred over this period, including in 2002, 2005, 2011, 2015, and 2017 (the 2017 failure was attributed to a programming error rather than the rocket).[65] The 2011 launch failure occurred on a Soyuz-U and was the first failure after 80 consecutive successful launches of the Soyuz-U and Soyuz-FG.[66] In April 2018, a Soyuz-FG rocket bound for the ISS was also forced to activate its launch abort system.[67] Despite these incidents, the Soyuz rocket maintains a reputation of very high reliability.

In 2018, 16 Soyuz were launched including 13 for Russia and 3 for Europe, capturing a global market share of 22 percent. Of all the Russian launches in 2018, 87 percent were launched on a Soyuz as only two Proton rockets were launched during the year.

---

[64] Some mission failures that occurred during re-entry, not during the launch period, are not considered failures of the Soyuz vehicle and are thus not within this scope. For example, the manned Soyuz 1 and Soyuz 11 missions in 1967 and 1971 ran into fatal problems during reentry. Howell, 2018.

[65] Howell, 2018 ; Anatoly Zak, "Soyuz Rocket Missions in 2017," *RussianSpaceWeb.com*, June 17, 2018.

[66] Lardier and Barensky, p. xiii.

[67] Matthew Bodner, "Soyuz Demonstrates Finesse in Flight and Failure," *SpaceNews*, October 11, 2018.

Table C.3 provides additional detail on the launches of the Soyuz family of launch vehicles as of January 2018. The first launch of the Soyuz 5 is planned for 2022.[68]

The Persistence of Arianespace

Arianespace was founded in 1980 as the first commercial launch service provider with the mission of "guaranteeing independent space access for Europe."[69] Arianespace has held a significant share of the commercial market since its creation, with more than 200 successful launches. Arianespace currently has three families of launch vehicles: the Ariane 5 is a heavy lift launch vehicle that entered service in 1998 and has completed 103 launches; the Soyuz is a medium lift launch vehicle with 47 launches since 2011; and the Vega is a small lift launch vehicle that entered service in 2012 and has completed 14 launches.[70] The recent diversification in Arianespace's launch vehicle portfolio indicates that the company might be trying to transition to a Strategy D provider as demonstrated in Figure C.8. As a subsidiary of ArianeGroup (a European aerospace conglomerate that includes Airbus), several multinational companies have capital in Arianespace, with French investors holding the largest stake at 64 percent.[71,72]

Table C.3
Launches of Soyuz Vehicles

Launch Vehicle	Year of First Launch	Number of Orbital Launches	Reliability
Soyuz FG	2001	62	100%
Soyuz-2.1a	2004	32	97%
Soyuz-2.1b	2006	37	94%
Soyuz-2.1v	2013	3	100%
Soyuz-5	2020	—	—

SOURCE: FAA, 2018.

---

[68] FAA, 2018.

[69] Arianespace, "Company Profile," webpage, undated-c.

[70] Arianespace, undated-c.

[71] Arianespace, undated-c.

[72] ArianeGroup is also a 51 percent shareholder of a joint venture called Eurockot Launch Service, with the 49 percent shareholder being Russia's Khrunichev State Research and Production Space Centre. Eurockot formed in 1995 and has executed more than 25 launches since 2000 using a vehicle called the Rockot. Caleb Henry, "Eurockot Conducts Final Rockot Mission with Sentinel-3B Satellite," *SpaceNews*, April 25, 2018c.

Figure C.8
Arianespace Begins to Use Strategy D, Diversifying Launch Vehicle Portfolio

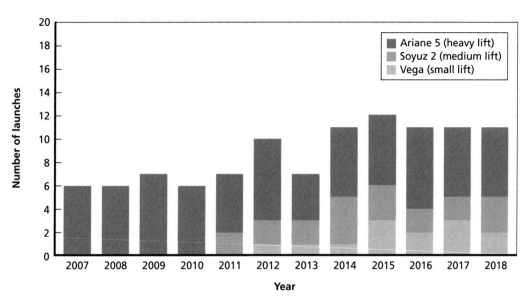

SOURCE: RAND analysis of heavy lift launch data set, 2007–2018 (Zak, 2019c; Union of Concerned Scientists, 2019; U.S. Department of Transportation, Bureau of Transportation Statistics, 2017; World Bank, undated; World Bank, 2018).

The company has launch facilities at the Guiana Space Center in French Guiana and additional launch facilities at Baikonur in Kazakhstan.[73]

Arianespace has been a major competitor in the heavy lift launch market for the past four decades, having held a market share similar to the other major launch service providers (Proton and Sea Launch) from 2000 to 2009. However, in recent years, Arianespace has slowly begun to lose market share. Taking decadal snapshots, Arianespace executed 20 percent of all heavy lift launches in 1998, 15 percent in 2008, and 12 percent in 2018.[74] The decline in Arianespace's market share is due, at least in part, to the rapid rise of SpaceX in the global launch market. The Falcon 9 entered competition in 2012 and by 2018 has captured 28 percent of the global market.

Concerns over Arianespace's future have emerged because of aggressive competition from SpaceX. In 2018, ASD-Eurospace, a nonprofit European space advocacy organization, released a report that expressed alarm regarding the resurgence of Ameri-

---

[73] Arianespace, undated-c.

[74] FAA, 1998; FAA, 2008; Stephen Clark, "Orbital ATK Confident New Rocket Will Win Air Force Support," *Spaceflight Now*, April 12, 2018a; Stephen Clark, "ULA, Blue Origin, Northrop Grumman Share Air Force Rocket Funding," *Spaceflight Now*, October 16, 2018b; Zak, 2018.

can dominance in space.[75] France's internal auditor, *Cour des comptes*, also released a 2019 report that noted SpaceX had "replaced Arianespace as the global leader on the commercial market."[76] In 2018, the chief executive officer of ArianeGroup told a German magazine that SpaceX was intentionally cutting prices to force Europe out of the market.[77]

Despite aggressive competition from SpaceX, Arianespace has remained competitive in the commercial launch market, executing six of the 25 addressable launches in 2018 and capturing the second largest market share. Moreover, Arianespace has taken steps in recent years to increase its competitiveness in the market. As of 2019, Arianespace is developing a next-generation heavy lift launch vehicle, the Ariane 6, that was "conceived for reduced production costs and design-to-build lead times, all while maintaining the quality and reliability that have made the Ariane 5 an industry leader," and is scheduled to become operational in 2020.[78] Arianespace has chosen not to develop the Ariane 6 as a reusable rocket (a method that has made SpaceX competitive). However, the Ariane 6 is expected to have twice the number of launches as the Ariane 5.[79] The Ariane 6 will also be capable of launching greater masses to orbit and allow dual manifesting of heavy payloads—a shortcoming of the Ariane 5, which must dual manifest heavy and lighter payloads. Arianespace hopes to reduce the cost of the Ariane 6 to approximately 40 percent of the Ariane 5 with these "design changes and higher volume production."[80]

Overall, Arianespace is positioning itself to regain market share through its efforts to develop new launch capabilities and reduce costs. Economic and political concerns regarding a single, dominant launch service provider will also likely allow Arianespace to regain at least some of its historical market share. However, a report from *Cour des comptes* claimed the model used to develop the Ariane 6 is "cautious" and "risks not being competitive over the long term."[81]

---

[75] Eurospace, "The Growing U.S. Ambition for Space Dominance, a New Challenge for European Independence in Space," *ASD-Eurospace* blog, October 12, 2018.

[76] Cour des comptes, "The Policy on Space Launchers: Significant Challenges to Overcome," in *2019 Annual Public Report, Volume 1: Observations*, Paris, February 6, 2019.

[77] Rich Smith, "Europe Complains: SpaceX Rocket Prices Are Too Cheap to Beat," *Motley Fool*, June 2, 2018; Christopher Seidler, "Die Amerikaner Wollen Europa aus dem Weltraum Kicken," *Spiegel Online*, May 18, 2018.

[78] Arianespace "Ariane 6," webpage, undated-b.

[79] Arthur Villasanta, "Arianespace vs. SpaceX: Ariane 6 Rockets Will Be 40% Cheaper Per Launch," *International Business Times*, February 1, 2019.

[80] Jamie Freed, "Europe's Arianespace Takes on SpaceX by Cutting Ariane 5 Rocket Launch Price," Reuters, January 23, 2019.

[81] As cited in Helene Fouquet, "Europe's Next-Generation Rocket is Doomed Even Before First Flight, Auditor Finds," *Bloomberg*, February 6, 2019. Cour des comptes, 2019.

Bibliography

ArianeSpace, "About Us," webpage, undated-a. As of August 1, 2019:
http://www.ArianeSpace.com/about-us/

———, "Ariane 6," webpage, undated-b. As of August 1, 2019:
http://www.ArianeSpace.com/ariane-6/

———, "Company Profile," webpage, undated-c. As of August 1, 2019:
http://www.ArianeSpace.com/company-profile/

———, "Launch Log," webpage, undated-d. As of August 1, 2019:
http://www.ArianeSpace.com/launch-log/

———, "The Spaceport," webpage, undated-e. As of August 1, 2019:
http://www.ArianeSpace.com/spaceport-facility/

———, "Technical Overview: Soyuz," webpage, undated-f. As of August 1, 2019:
http://www.ArianeSpace.com/wp-content/uploads/2019/04/ARIANESPACE-ENG-FLYER-
SOYUZ-APRIL2019-WEB.pdf

———, "Satellite 2019: Ariane 6 and Vega C Are Coming! ArianeSpace Places its Operational
Excellence and Commercial Performance at the Service of Clients," webpage, May 6, 2019. As of
August 1, 2019:
http://www.ArianeSpace.com/press-release/satellite-2019/

"Atlas ICBM Chronology," webpage, last updated February 2006. As of August 1, 2019:
https://web.archive.org/web/20060204073649/http://www.geocities.com/atlas_missile/Chronology.
html#expand

Aviation Week and Space Technology, "First RD-180 Delivered," March 24, 1997, p. 27.

———, "The U.S. Federal Trade Commission Has Approved the Merger of Lockheed Martin's and
Boeing's Government Launch Business," October 9, 2006, p. 21.

———, "SpaceX Wins NASA Launch Services Contract," April 24, 2008.

Bergin, Chris, "SpaceX and Orbital Win Huge CRS Contract from NASA," *NASASpaceFlight.com*,
December 23, 2008. As of October 2, 2019:
https://www.nasaspaceflight.com/2008/12/spacex-and-orbital-win-huge-crs-contract-from-nasa/

———, "Space Industry Giants Orbital Upbeat Ahead of Antares Debut," *NASASpaceFlight.com*,
February 22, 2012a. As of October 2, 2019:
https://www.nasaspaceflight.com/2012/02/orbital-upbeat-ahead-of-antares-debut/

———, "Orbital's Cygnus Debut Mission to the ISS Outlined," *NASASpaceFlight.com*, June 4,
2012b. As of October 2, 2019:
https://www.nasaspaceflight.com/2012/06/orbitals-cygnus-debut-mission-iss-outlined/

Broad, William J., "Pentagon Leaves the Shuttle Program," *New York Times*, August 7, 1986. As of August 1, 2019:
https://www.nytimes.com/1989/08/07/us/pentagon-leaves-the-shuttle-program.html

Bodner, Matthew, "Soyuz Demonstrates Finesse in Flight and Failure," *SpaceNews*, October 11, 2018. As of August 1, 2019:
https://spacenews.com/soyuz-demonstrates-finesse-in-flight-and-failure/

Boeing, "U.S. Air Force Procures Boeing Delta IV Launches for EELV Program," press release, October 16, 1998. As of October 2, 2019:
https://boeing.mediaroom.com/1998-10-16-U.S.-Air-Force-Procures-Boeing-Delta-IV-Launches-for-EELV-Program

Bunko, Tracy, "New Entrant Certification Strategy Announced," Washington, D.C., Secretary of the Air Force Public Affairs, U.S. Air Force, October 14, 2011. As of October 1, 2019:
https://www.af.mil/News/Article-Display/Article/112266/
new-entrant-certification-strategy-announced/

Butler, Amy, "Playing the Long Game, Legal Settlement Offers SpaceX Little Near-term Gain, but it Serves the Company's Strategic Goals," *Aviation Week and Space Technology*, February 2, 2015a, pp. 33–34.

———, "Soft Launch, Hard Impact, Falcon 9 Certification is a Baseline for Future Upgrades to SpaceX Launch Vehicles," *Aviation Week and Space Technology*, June 8, 2015b, p. 56.

———, "Grounded, Government Intervention in RD-180 Supply Hampers ULA's Competitiveness, CEO Says," *Aviation Week and Space Technology*, October 12, 2015c, pp. 34–36.

Carreau, Mark, "Up and Back SpaceX Opens the Hatch to Commercial Resupply," *Aviation Week and Space Technology*, June 4, 2012, pp. 36-37.

Clark, Stephen, "Orbital ATK Confident New Rocket Will Win Air Force Support," *Spaceflight Now*, April 12, 2018a. As of August 1, 2019:
https://spaceflightnow.com/2018/04/12/orbital-atk-confident-new-rocket-will-win-air-force-support/

———, "ULA, Blue Origin, Northrop Grumman Share Air Force Rocket Funding," *Spaceflight Now*, October 16, 2018b. As of August 1, 2019:
https://spaceflightnow.com/2018/10/16/ula-blue-origin-northrop-grumman-share-air-force-rocket-funding/

———, "Viasat Swaps Ariane 5 Launch for New Ariane 6 Rocket," *Spaceflight Now*, June 17, 2019. As of August 1, 2019:
https://spaceflightnow.com/2019/06/17/viasat-swaps-ariane-5-launch-for-new-ariane-6-rocket/

Comptroller General of the United States, *Cost Benefit Analysis Used in Support of the Space Shuttle Program: Report to Congress*, Washington, D.C.: U.S. Government Accountability Office, B-173777, June 2, 1972. As of August 1, 2019:
http://archive.gao.gov/f0302/096542.pdf

Connected-Earth, "The First Satellites," webpage, undated. As of August 1, 2019:
http://www.connected-earth.com/Journeys/Firstgenerationtechnologies/Satelliteandmicrowave/Thefirstsatellites/index.htm

Cooper, John Cobb, "The Manned Orbiting Laboratory: A Major Legal and Political Decision," *ABAJ*, Vol. 51, December 1965, p. 1137.

Cour des comptes, "The Policy on Space Launchers: Significant Challenges to Overcome," in *2019 Annual Public Report, Volume 1: Observations*, Paris, February 6, 2019. As of July 1, 2019:
https://www.ccomptes.fr/system/files/2019-06/Policy-space-launchers-vol-1.pdf

Cox, Christopher, and the Select Committee of the United States House of Representatives, "U.S. Export Policy Toward the PRC," in *Report of the Select Committee on U.S. National Security and Military/Commercial Concerns with the People's Republic of China*, Washington, D.C.: U.S. Government Publishing Office, May 25, 1999. As of August 1, 2019: https://www.govinfo.gov/content/pkg/GPO-CRPT-105hrpt851/pdf/GPO-CRPT-105hrpt851-3-5.pdf

Cox, Louis A., Jr., "What's Wrong With Risk Matrices?" *Risk Analysis*, Vol. 28, No. 2, April 2008, pp. 497–512.

Dean, James, "Air Force Awards Major Contracts to ULA, Northrop Grumman and Blue Origin," *Florida Today*, October 10, 2018. As of August 1, 2019: https://eu.floridatoday.com/story/tech/science/space/2018/10/10/air-force-awards-major-contracts-ula-northrop-and-blue-origin/1594594002/

de Carbonnel, Alissa, "Russia Bets on Sweeping Reform to Revive Ailing Space Industry," Reuters, December 26, 2013. As of July 19, 2019: https://www.reuters.com/article/us-russia-space/russia-bets-on-sweeping-reform-to-revive-ailing-space-industry-idUSBRE9BP02S20131226

Department of the Air Force, Air Force Space Command, "FA8811-17-9-0001; Evolved Expendable Launch Vehicle (EELV) Launch Service Agreements (LSA) Request for Proposals (RFP)," Solicitation Number FA8811-16-R-000X, March 2017. As of October 1, 2019: https://www.fbo.gov/index?s=opportunity&mode=form&id=f951493909b0042981ef23bbaec26bcf&tab=core&_cview=1

de Selding, Peter B., "Sea Launch Rocket Failure Destroys Intelsat IS-27 Satellite," *SpaceNews*, February 1, 2013. As of May 1, 2019: https://spacenews.com/33452sea-launch-rocket-failure-destroys-intelsat-is-27-satellite/

———, "SpaceX Aims to Debut New Version of Falcon 9 This Summer," *SpaceNews*, March 20, 2015. As of October 2, 2019: https://spacenews.com/spacex-aims-to-debut-new-version-of-falcon-9-this-summer/

Deutsche Welle, "ArianeSpace's Vega Rocket Fails for First Time," *DW* blog, July 11, 2019. As of July 11, 2019: https://www.dw.com/en/ArianeSpaces-vega-rocket-fails-for-the-first-time/a-49545126-0

DiMascio, Jen, "The Next Generation of GPS Satellites," *Aviation Week and Space Technology*, December 24, 2018, p. 12.

Dornheim, Michael A., "Back to Basics Government Abandons Commercial Practices on EELV," *Aviation Week and Space Technology*, March 13, 2006, p. 28. As of August 1, 2019: http://archive.aviationweek.com/search?QueryTerm=ELC

Dubois, Thierry, "ArianeSpace Needs Both Commercial And Institutional Orders," *Aviation Week and Space Technology*, January 16, 2019. As of August 1, 2019: https://aviationweek.com/space/ArianeSpace-needs-both-commercial-and-institutional-orders

Encyclopedia Astronautica, "Delta D," webpage, undated-a. As of August 1, 2019: http://www.astronautix.com/d/deltad.html

———, "Thor," webpage, undated-b. As of August 1, 2019: http://www.astronautix.com/t/thor.html

———, "Delta," webpage, last updated 2008. As of August 1, 2019: https://web.archive.org/web/20080329150203/http://www.astronautix.com/lvfam/delta.htm

Erwin, Sandra, "Air Force to Continue to Push Back on Proposed Space Launch Legislation," *SpaceNews*, June 27, 2019. As of September 1, 2019:
https://spacenews.com/air-force-to-continue-to-push-back-on-proposed-space-launch-legislation/

———, "For OmegA, U.S. Air Force Launch Competition Is a Must-win," *SpaceNews*, April 8, 2019. As of May 1, 2019:
https://spacenews.com/for-omega-u-s-air-force-launch-competition-is-a-must-win/

European Space Agency, "Soyuz," webpage, undated. As of June 2019:
https://www.esa.int/Our_Activities/Space_Transportation/Launch_vehicles/Soyuz

———, "Soyuz Launcher," *Human Spaceflight* blog, May 2005. As of August 1, 2019:
http://www.spaceflight.esa.int/documents/foton/spacecraft.pdf

———, *ESA Annual Report*, The Netherlands, AR-2016, December 2016. As of August 1, 2019:
https://www.esa.int/About_Us/ESA_Publications/ESA_Publications_Annual_Report/ESA_Annual_Report_2016

———, "Launch Vehicles: Soyuz MS Spacecraft," webpage, December 20, 2018. As of August 1, 2019:
https://www.esa.int/Our_Activities/Space_Transportation/Launch_vehicles/The_Russian_Soyuz_spacecraft

Eurospace, "The Growing US Ambition for Space Dominance, a New Challenge for European Independence in Space," *ASD-Eurospace* blog, October 12, 2018. As of August 1, 2019:
https://eurospace.org/eurospace-position-paper-the-growing-us-ambition-for-space-dominance-a-new-challenge-for-european-independence-in-space/

Federal Aviation Administration, *Commercial Space Transportation: 1998 Year in Review*, Washington, D.C., January 1999. As of October 2, 2019:
https://www.faa.gov/about/office_org/headquarters_offices/ast/media/1998yir.pdf

———, *Commercial Space Transportation: 2008 Year in Review*, Washington, D.C., January 2009. As of October 2, 2019:
https://www.faa.gov/about/office_org/headquarters_offices/ast/media/2008%20Year%20in%20Review.pdf

———, *Semi-Annual Launch Report: Second Half of 2009*, Washington, D.C., 2009.

———, *Commercial Space Transportation: 2011 Year in Review*, Washington, D.C., January 2012.

———, *The Annual Compendium of Commercial Space Transportation: 2012*, Washington, D.C., February 2013.

———, *The Annual Compendium of Commercial Space Transportation: 2017*, Washington, D.C., January 2017.

———, *The Annual Compendium of Commercial Space Transportation: 2018*, Washington, D.C., 2018. As of August 1, 2019:
https://www.faa.gov/about/office_org/headquarters_offices/ast/media/2018_ast_compendium.pdf

Foust, Jeff, "First Falcon Heavy Launch Scheduled for Spring," *SpaceNews*, September 2, 2015. As of October 2, 2019:
https://spacenews.com/first-falcon-heavy-launch-scheduled-for-spring/

———, "Air Force Adds More Than $40 Million to SpaceX Engine Contract," *SpaceNews*, October 21, 2017. As of October 2, 2019:
https://spacenews.com/air-force-adds-more-than-40-million-to-spacex-engine-contract/

————, "Do Smallsats Even Need Insurance?" *SpaceNews*, October 4, 2018. As of October 2, 2019:
https://spacenews.com/do-smallsats-even-need-insurance/

————, "House Appropriators Take a Pass on NASA Budget Amendment," *SpaceNews*, May 16, 2019. As of August 1, 2019:
https://spacenews.com/house-appropriators-pass-on-nasa-budget-amendment/

Fouquet, Helene, "Europe's Next-Generation Rocket is Doomed Even Before First Flight, Auditor Finds," *Bloomberg*, February 6, 2019. As of August 1, 2019:
https://www.bloomberg.com/news/articles/2019-02-06/
before-even-launching-ariane-6-rocket-journey-is-seen-as-doomed

Freed, Jamie, "Europe's ArianeSpace Takes on SpaceX by Cutting Ariane 5 Rocket Launch Price," Reuters, January 23, 2019. As of August 1, 2019:
https://www.reuters.com/article/arianespace-asia/
europes-arianespace-takes-on-spacex-by-cutting-ariane-5-rocket-launch-price-idUSL3N1ZN2JF

Furniss, Tim, "Launchers Directory," *Flight International*, December 11–17, 1996. As of October 2, 2019:
https://www.flightglobal.com/FlightPDFArchive/1996/1996%20-%203297.PDF

Gainor, Christopher, "The Atlas and the Air Force: Reassessing the Beginnings of America's First Intercontinental Ballistic Missile," *Technology and Culture*, Vol. 54, No. 2, April 2013, pp. 346–370.

Gallois, Dominique, "Ariane 6, un Chantier Européen pour Rester dans la Course Spatiale," *Le Monde*, December 2, 2014. As of August 1, 2019:
http://www.lemonde.fr/economie/article/2014/12/01/les-europeens-s-appretent-a-mettre-ariane-6-en-chantier_4532259_3234.html

GlobalSecurity.org, "Evolved Expendable Launch Vehicle (EELV) Program Overview," webpage, November 20, 1997. As of October 2, 2019:
https://www.globalsecurity.org/space/library/report/1997/nov_ovrw.pdf

Governments of the United Kingdom of Great Britain and Northern Ireland, the Union of Soviet Socialist Republics and the United States of America, "Treaty on Principles Governing the Activities of States in the Exploration and Use of Outer Space, including the Moon and Other Celestial Bodies," New York: United Nations General Assembly, January 27, 1967. As of October 1, 2019:
http://www.unoosa.org/oosa/en/ourwork/spacelaw/treaties/introouterspacetreaty.html

Government of India, "ISRO Budget at a Glance," webpage, undated. As of June 6, 2019:
https://www.isro.gov.in/budget-glance

Gunter's Space Page, homepage, undated. As of October 11, 2019:
https://space.skyrocket.de/

Henry, Caleb, "Russia Merges United Rocket and Space Corporation with Roscosmos," *SatelliteToday*, January 23, 2015. As of July 19, 2019:
https://www.satellitetoday.com/business/2015/01/23/
russia-merges-united-rocket-and-space-corporation-with-roscosmos/

————, "SpaceX's Final Falcon 9 Design Coming This Year, two Falcon Heavy Launches Next Year," *SpaceNews*, June 27, 2017. As of October 2, 2019:
https://spacenews.com/spacexs-final-falcon-9-design-coming-this-year-two-falcon-heavy-launches-next-year/

————, "Sea Launch CEO Sergey Gugkaev to Leave Company when S7 Purchase Closes," *SpaceNews*, March 13, 2018a. As of August 1, 2019:
https://spacenews.com/sea-launch-ceo-sergey-gugkaev-to-leave-company-when-s7-purchase-closes/

————,"S7 Closes Sea Launch Purchase, Future Rocket TBD," *SpaceNews*, April 17, 2018b. As of August 1, 2019:
https://spacenews.com/s7-closes-sea-launch-purchase-future-rocket-tbd/

————,"Eurockot Conducts Final Rockot Mission with Sentinel-3B Satellite," *SpaceNews*, April 25, 2018c. As of August 1, 2019:
https://spacenews.com/eurockot-conducts-final-rockot-mission-with-sentinel-3b-satellite/

————, "SpaceX to Launch 'Dozens' of Starlink Satellites Next Week, More to Follow," *SpaceNews*, May 8, 2019a. As of June 2019:
https://www.space.com/spacex-starlink-satellites-launching-may-2019.html

————, "SpaceX Targets 2021 Commercial Starship Launch," *SpaceNews*, June 28, 2019b. As of October 1, 2019:
https://spacenews.com/spacex-targets-2021-commercial-starship-launch/

Heydon, Douglas, "ArianeSpace—Risks and Rewards in the Launch Service Business," paper presented at 23rd Joint Propulsion Conference, San Diego, Calif.: American Institute of Aeronautics and Astronautics, June 29–July 2, 1987. As of October 1, 2019:
https://arc.aiaa.org/doi/10.2514/6.1987-1797

Howell, Elizabeth, "Soyuz Rocket: Russia's Venerable Booster," *Space.com*, April 12, 2018. As of August 1, 2019:
https://www.space.com/40282-soyuz-rocket.html

ILS—*See* International Launch Services.

International Launch Services, "About Us," webpage, undated. As of July 19, 2019:
https://www.ilslaunch.com/about-us/

————, "Creation of International Launch Services," webpage, June 10, 1995. As of July 19, 2019:
https://www.ilslaunch.com/creation-of-international-launch-services/

————, "ILS Opens New Era with New Launch Pricing; Will Operate Under Glavkosmos," webpage, April 12, 2019a. As of July 19, 2019:
https://www.ilslaunch.com/ils-opens-new-era-with-new-launch-pricing-will-operate-under-glavkosmos/

————, "Proton-M Successfully Launches Spektr-RG; 5-Year Reliability Rises to 95.7%," webpage, July 15, 2019b. As of July 19, 2019:
https://www.ilslaunch.com/proton-with-spektr-rg-observatory-successfully-launched-from-baikonur/

"India Approves US$1.574 Billion Funding for ISRO's 30 PSLV & 10 GSLV Flights," Spacetech, June 6, 2018. As of August 1, 2019:
http://www.spacetechasia.com/india-approves-us1-574-billion-funding-for-30-pslv-10-gslv-flights/

JAXA—*See* Japan Aerospace Exploration Agency.

Japan Aerospace Exploration Agency, "About H-IIB Launch Vehicle," webpage, undated-a. As of August 1, 2019: https://global.jaxa.jp/projects/rockets/h2b/index.html

————, "H-IIA Launch Vehicle," webpage, undated-b. As of August 1, 2019:
https://global.jaxa.jp/projects/rockets/h2a/

Jones, Harry W., *The Recent Large Reduction in Space Launch Cost*, Albuquerque, New Mexico: 48th International Conference on Environmental Systems, ICES-2018-81, July 8–12, 2018.

Jurkowsky, Tom, "Lockheed Martin Announces Sale of its Interests in International Launch Services and LKEI," Lockheed Martin, September 7, 2006, As of August 1, 2019:
https://web.archive.org/web/20081007044732/http://www.lockheedmartin.com/news/press_releases/2006/LockheedMartinAnnouncesSaleOfItsInt.html

Karash, Yuri, "Russian Space Program: Financial State, Current Plans, Ambitions and Cooperation with the United States," Cape Canaveral, Fla.: *Space Congress Proceedings*, Vol. 27, May 26, 2016. As of June 1, 2019:
https://commons.erau.edu/cgi/viewcontent.cgi?article=3648&context=space-congress-proceedings

Klotz, Irene, "Game On," *Aviation Week and Space Technology*, April 9, 2018, pp. 44–47.

Klotz, Irene, and Jen DiMascio, "SpaceX Loses Out on U.S. Air Force Next-Gen Launcher Development," *Aviation Week and Space Technology*, October 15, 2018, p. 38.

Koren, Marina, "The Fraught Effort to Return to the Moon," *The Atlantic*, July 17, 2019.

Korn, Paula, and Julie Fornaro, "One World, One Platform: For Sea Launch's Multinational Crew, a Cruise on the Pacific Ocean Means Getting Ready for a Rocket's Liftoff," *Boeing Frontiers*, Vol. 1, No. 2, 2002.

Khrunichev State Research and Production Space Center, "Angara Launch Vehicles Family," webpage, undated. As of August 1, 2019:
http://www.khrunichev.ru/main.php?id=44&hl=angara

Lardier, Christian, and Stefan Barensky, *The Soyuz Launch Vehicle: The Two Lives of an Engineering Triumph*, New York: Springer Science+Business Media, 2013.

Lee, Ming-Chang, To Chang, and Wen-Tien Chang Chien, "An Approach for Developing Concept of Innovation Readiness Levels," *International Journal of Managing Information Technology*, Vol. 3, No. 2, May 2011. As of October 1, 2019:
https://pdfs.semanticscholar.org/fabf/066f571a5b7572b0d4c69755c00a251323e4.pdf

Levine, E.S., "Improving Risk Matrices: The Advantages of Logarithmically Scaled Axes," *Journal of Risk Research*, Vol. 15, No. 2, February 2012, pp. 209–222.

Lockheed Martin Space Systems Company, "Lockheed Martin's Last Titan IV Successfully Delivers National Security Payload to Space," press release, October 19, 2005. As of August 1, 2019:
https://news.lockheedmartin.com/2005-10-19-Lockheed-Martins-Last-Titan-IV-Successfully-Delivers-National-Security-Payload-to-Space

Logsdon, John M., ed., *Exploring The Unknown: Selected Documents in the History of the U.S. Civil Space Program Volume IV: Accessing Space*, Washington, D.C.: National Aeronautics and Space Administration History Division, Office of Policy and Plans, 1999.

———, "Launch Vehicle," *Encyclopedia Britannica*, March 13, 2019. As of August 1, 2019:
https://www.britannica.com/technology/launch-vehicle

McCartney, Forrest, Peter A. Wilson, Lyle Bien, Thor Hogan, Leslie Lewis, Chet Whitehair, Delma Freeman, T. K. Mattingly, Robert Larned, David S. Ortiz, William A. Williams, Charles J. Bushman, and Jimmey Morrell, *National Security Space Launch Report*, Santa Monica, Calif.: RAND Corporation, MG-503-OSD, 2006. As of July 20, 2019:
https://www.rand.org/pubs/monographs/MG503.html

Mitsubishi Heavy Industries, *H-IIA User's Manual, Version 4.0*, Tokyo, Japan: Mitsubishi, February 2015. As of October 1, 2019:
https://www.mhi.com/jp/products/pdf/manual.pdf

Morring, Frank, Jr., "Five Vehicles Vie To Succeed Space Shuttle," *Aviation Week and Space Technology*, April 29, 2011.

———, "In Orbit: United Alliance Launch," *Aviation Week and Space Technology*, December 11, 2006, p. 17. As of October 1, 2019:
https://archive.aviationweek.com/issue/20061211

Narasimhan, T.E., "'Fat Boy' GSLV-MK III Launches Today: The Rocket Has Cost India Rs 400 cr," *Business Standard*, June 5, 2017. As of June 6, 2019:
https://www.business-standard.com/article/current-affairs/fat-boy-gslv-mk-iii-launches-today-the-rocket-has-cost-india-rs-400-cr-117060500106_1.html

NASA—*See* National Aeronautics and Space Administration.

National Aeronautics and Space Administration, "Syncom 2," Washington, D.C., NASA Space Science Data Coordinated Archive/COSPAR ID: 1963-031A, undated.

———, "Kennedy Space Center Launch Archives: 1981–1986 Space Shuttle Launches," webpage, last updated February 24, 2008. As of October 1, 2019:
https://www.nasa.gov/centers/kennedy/shuttleoperations/archives/1981-1986.html

———, Blue Origin Space Agreement, Washington, D.C.: NASA Crew and Cargo Program Office, 2010.

———, "NASA Launch Services Manifest," July 1, 2011. As of October 2, 2019:
https://www.nasa.gov/pdf/315550main_launch_manifest_07_01_2011.pdf

———, "Commercial Orbital Transportation Services a New Era in Spaceflight," 2014.

———, *National Aeronautics and Space Administration FY 2018 Agency Financial Report*, Washington, D.C, 2018.

National Research Council, *Pathways to Exploration: Rationales and Approaches for a U.S. Program of Human Space Exploration*, Washington, D.C.: National Academies Press, 2014. As of August 1, 2019:
https://www.nap.edu/catalog/18801/pathways-to-exploration-rationales-and-approaches-for-a-us-program

Northrop Grumman, "OmegA—The Smart Choice," webpage, undated. As of October 1, 2019:
http://www.northropgrumman.com/MediaResources/MediaKits/OmegARocket/Home.aspx

Nowakowski, Tomasz, "Russia Dissolves Its Federal Space Agency, What Now?" *SpaceFlight Insider*, December 30, 2015. As of July 19, 2019:
https://www.spaceflightinsider.com/organizations/roscosmos/russia-dissolves-federal-space-agency-now/

Oberhaus, Daniel, "SpaceX Is Banking on Satellite Internet. Maybe It Shouldn't," *Wired*, May 19, 2019.

O'Callaghan, Jonathan, "'Not Good Enough'—SpaceX Reveals That 5% Of Its Starlink Satellites Have Failed In Orbit So Far," *Forbes*, June 30, 2019. As of October 2, 2019:
https://www.forbes.com/sites/jonathanocallaghan/2019/06/30/not-good-enough-spacex-reveals-that-5-of-its-starlink-satellites-have-failed-in-orbit/#484f00287e6b

Pasztor, Andy, "Startup Satellite Venture OneWeb Blasts Off With Revised Business Plan," *Wall Street Journal*, February 27, 2019.

Patterson, Trina, "Northrop Grumman OmegA Rocket Team Celebrates Air Force Launch Service Agreement Award," Northrop Grumman, October 12, 2018. As of August 1, 2019:
https://news.northropgrumman.com/news/features/northrop-grumman-omega-rocket-team-celebrates-air-force-launch-service-agreement-award

Perrett, Bradley, "Chinese Working On Giant Engine For Long March 9," *Aviation Week and Space Technology*, March 12, 2018a. As of August 1, 2019:
https://aviationweek.com/space/chinese-working-giant-engine-long-march-9

———, "Long March 8 Reconfigured For Reuse," *Aviation Week and Space Technology*, October 26, 2018b. As of August 1, 2019:
https://aviationweek.com/space/long-march-8-reconfigured-reuse

———, "Second Version Of Long March 7 Awaits Approval," *Aerospace Daily and Defense Report*, October 26, 2018c. As of August 1, 2019:
https://aviationweek.com/awinspace/second-version-long-march-7-awaits-approval

Plass, Simon, Federico Clazzer, and Fritz Bekkadal, *Current Situation and Future Innovations in Arctic Communications*, in 2015 IEEE 82nd Vehicular Technology Conference (VTC2015-Fall), Boston: Institute of Electrical and Electronic Engineers, 2015, pp. 1–7. As of October 2, 2019:
https://ieeexplore.ieee.org/document/7390883

Presidential Commission on the Space Shuttle Challenger Accident, *Report to the President by the Presidential Commission on the Space Shuttle Challenger Accident*, Washington, D.C., June 6, 1986.

Public Law 114-328, National Defense Authorization Act for Fiscal Year 2017, Section 1602, (c) (2), Exception to the Prohibition on Contracting with Russian Suppliers of Rocket Engines for the Evolved Expendable Launch Vehicle Program, December 23, 2016. As of October 2, 2019:
https://www.congress.gov/114/plaws/publ328/PLAW-114publ328.pdf

Pultarova, Tereza, "Largest All-Electric Satellite to Date Completes Orbit Raising in Record Time," *SpaceNews*, October 12, 2017.

Rhian, Jason, "Next Generation Launcher Considered Under U.S. Air Force's EELV Program," *SpaceFlight Insider*, March 24, 2018. As of August 1, 2019:
https://www.spaceflightinsider.com/missions/defense/
next-generation-launcher-considered-u-s-air-forces-eelv-program/

Ruffner, Kevin C., ed., *CORONA: America's First Satellite Program*, Washington, D.C.: Central Intelligence Agency, Center for the Study of Intelligence, 1995. As of August 1, 2019:
https://www.cia.gov/library/center-for-the-study-of-intelligence/csi-publications/books-and-monographs/corona.pdf

"Russia to Allocate $1.5Bln to Federal Space Program in 2017—Draft Budget Plan," *Sputnik News*, October 29, 2016. As of August 1, 2019:
https://sputniknews.com/russia/201610291046871259-money-program-space-russia/

S7 Space, "Sea Launch," webpage, undated. As of August 1, 2019:
http://s7space.ru/en/launch-sea/

Scully, Janene, "New Era Dawns for SLC-6," *Lompoc Record*, September 29, 2005. As of October 1, 2019:
https://lompocrecord.com/news/local/new-era-dawns-for-slc/article_bd0a0f6c-85c7-5d57-b11a-83523377aca9.html

Seidler, Christoph, "'Die Amerikaner Wollen Europa aus dem Weltraum Kicken,'" *Spiegel Online*, May 18, 2018. As of August 1, 2019:
https://www.spiegel.de/wissenschaft/technik/alain-charmeau-die-amerikaner-wollen-europa-aus-dem-weltraum-kicken-a-1207322.html

Sharkey, Jim, "Sea Launch Reduces Staff Due to Lull in Launches," *SpaceFlight Insider*, August 27, 2014. As of August 1, 2019:
https://www.spaceflightinsider.com/space-flight-news/sea-launch-reduces-staff-due-lull-launches/

Smith, Rich, "Europe Complains: SpaceX Rocket Prices Are too Cheap to Beat," *Motley Fool*, June 2, 2018. As of August 1, 2019:
https://www.fool.com/investing/2018/06/02/europe-complains-spacex-rocket-prices-are-too-chea.aspx

Spaceflight Now, homepage, undated. As of October 11, 2019:
https://spaceflightnow.com/

Space Launch Report, "LVM3 (GSLV Mk 3)," webpage, last updated July 22, 2019a. As of October 1, 2019:
http://www.spacelaunchreport.com/gslvmk3.html

———, "Proton Data Sheet," webpage, last updated August 5, 2019b. As of October 1, 2019:
http://www.spacelaunchreport.com/proton.html

———, "R-7/Soyuz Data Sheet," webpage, last updated September 26, 2019c. As of October 1, 2019:
https://www.spacelaunchreport.com/soyuz.html

Space and Missile Systems Center, "Launch Enterprise Systems Directorate, Fiscal Year 2020 President's Budget," May 29, 2019.

———, "Air Force Awards Final Rocket Propulsion System Prototype OTAs," 2016a.

———, "Air Force Awards Two Rocket Propulsion System Prototype OTAs," 2016b.

SpaceX, "SpaceX Announces Launch Date for the World's Most Powerful Rocket," press release, April 5, 2011. As of October 2, 2019:
https://www.spacex.com/press/2012/12/19/spacex-announces-launch-date-worlds-most-powerful-rocket

Stevenson, Richard W., "Shaky Start for Rocket Business," *New York Times*, September 16, 1988. As of August 1, 2019:
https://www.nytimes.com/1988/09/16/business/shaky-start-for-rocket-business.html

Svitak, Amy, "Fingers Crossed," *Aviation Week and Space Technology*, October 14, 2013, pp. 37–38.

ULA—*See* United Launch Alliance.

"ULA Team Launches USAF's AFPSC-11 Multi-Payload Mission," *Air Force Technology*, April 2018. As of October 2, 2019:
https://www.airforce-technology.com/news/ula-team-launches-usafs-afspc-11-multi-payload-mission/

"ULA Deal Cleared, but Not Final," *SpaceNews*, June 29, 2004. As of October 11, 2019:
https://spacenews.com/ula-deal-cleared-not-final/

Union of Concerned Scientists, "UCS Satellite Database," webpage, last updated March 31, 2019. As of October 11, 2019:
https://www.ucsusa.org/resources/satellite-database

United Launch Alliance, *Atlas V Launch Services User's Guide March 2010*, Littleton, Colo.: Lockheed Martin, 2010. As of June 11, 2019:
https://www.slideshare.net/BenjaminSpencerBilge/atlas-v-usersguide2010

United Nations, "Application of the Concept of the 'Launching State,'" New York, UN Resolution A/RES/59/115, January 25, 2005. As of August 1, 2019:
http://www.unoosa.org/pdf/gares/ARES_59_115E.pdf

USAF—*See* U.S. Air Force.

U.S. Air Force, *Launch Services New Entrant Certification Guide*, Washington, D.C.: U.S. Accountability Office, GAO 13-317R, February 2013. As of October 2, 2019:
https://apps.dtic.mil/dtic/tr/fulltext/u2/a580766.pdf

U.S. Department of Transportation, Bureau of Transportation Statistics, "Table 1-39: Worldwide Commercial Space Launches," webpage, last updated May 21, 2017. As of October 11, 2019:
https://www.bts.gov/archive/publications/national_transportation_statistics/table_01_39

U.S. Code, Title 10, Section 2273, Policy Regarding Assured Access to Space: National Security Payloads, in effect as of January 7, 2011.

U.S. Government Accountability Office, *Evolved Expendable Launch Vehicle: The Air Force Needs to Adopt an Incremental Approach to Future Acquisition Planning to Enable Incorporation of Lessons Learned*, Washington, D.C., GAO 15-623, August 2015. As of October 2, 2019:
https://www.gao.gov/assets/680/671926.pdf

———, *NASA: Commercial Partners Are Making Progress, but Face Aggressive Schedules to Demonstrate Critical Space Station Cargo Transport Capabilities*, GAO-09-618, Washington, D.C., June 16, 2009.

———, *NASA: Medium Launch Transition Strategy Leverages Ongoing Investments but Is Not Without Risk*, GAO-11-107, Washington, D.C., November 22, 2010. As of October 2, 2019:
https://www.gao.gov/assets/320/312661.html

———, *Evolved Expendable Launch Vehicle: DOD Is Assessing Data on Worldwide Launch Market to Inform New Acquisition Strategy*, GAO-16-661R, Washington, D.C., July 22, 2016. As of August 1, 2019:
https://www.gao.gov/assets/680/678646.pdf

———, *NASA Commercial Crew Program: Schedule Pressure Increases as Contractors Delay Key Events*, GAO-17-137, Washington, D.C., February 2017a. As of October 2, 2019:
https://www.gao.gov/assets/690/682859.pdf

———, *Surplus Missile Motors: Sale Price Drives Potential Effects on DOD and Commercial Launch Providers*, GAO-17-609, Washington, D.C., August 2017b. As of August 1, 2019:
https://www.gao.gov/assets/690/687571.pdf

———, *Space Acquisitions: DOD Faces Significant Challenges as it Seeks to Address Threats and Accelerate Space Programs*, testimony before the Subcommittee on Strategic Forces, Committee on Armed Forces, House of Representatives, Washington, D.C., April 3, 2019. As of October 2, 2019:
https://www.gao.gov/assets/700/698194.pdf

Vandenberg Air Force Base, "Evolved Expendable Launch Vehicle (EELV)," fact sheet, August 4, 2017. As of October 2, 2019:
https://www.vandenberg.af.mil/About-Us/Fact-Sheets/Display/Article/1266632/
evolved-expendable-launch-vehicle-eelv/

Villasanta, Arthur, "ArianeSpace vs. SpaceX: Ariane 6 Rockets Will Be 40% Cheaper Per Launch," *International Business Times*, February 1, 2019. As of August 1, 2019:
https://www.ibtimes.com/ArianeSpace-vs-spacex-ariane-6-rockets-will-be-40-cheaper-
launch-2758492

Wall, Mike, "More Power! SpaceX's Rockets Are Stronger Than Predicted," *Space.com*, May 2, 2016. As of October 2, 2019:
https://www.space.com/32767-spacex-rockets-falcon-9-capabilities.html

Waterman, Shaun, "Panelists See Launch Services Market, Providers, Contracting," Washington, D.C., *SATELLITE 2019*, May 8, 2019. As of August 1, 2019:
http://interactive.satellitetoday.com/via/satellite-2019-show-daily-day-3/
panelists-see-launch-services-market-providers-contracting/

"What Happened to Sea Launch," *Space Daily*, September 7, 2016. As of August 1, 2019:
http://www.spacedaily.com/reports/What_Happened_to_Sea_Launch_999.html

World Bank, "World Bank Country and Lending Gap," webpage, undated. As of October 11, 2019:
https://datahelpdesk.worldbank.org/knowledgebase/
articles/906519-world-bank-country-and-lending-groups

———, "Classifying Countries by Income," webpage, October 4, 2018. As of August 28, 2019:
http://datatopics.worldbank.org/world-development-indicators/stories/the-classification-of-countries-by-income.html

Yukman, Becki, "International Space Budget Data," in *The Space Report*, Colorado Springs, Colo.: Space Foundation, 2019. As of August 1, 2019:
https://promo.spacefoundation.org/tsr/quarterly-reports/TheSpaceReport19Q1.pdf

Zak, Anatoly, "Sea Launch Venture," *RussianSpaceWeb.com*, February 4, 2007. As of August 1, 2019:
http://www.russianspaceweb.com/sealaunch.html

———, "Soyuz Rocket Missions in 2017," *RussianSpaceWeb.com*, June 17, 2018. As of August 1, 2019:
http://www.russianspaceweb.com/soyuz_lv_2017.html

———, "Angara-5 to Replace Proton," *RussianSpaceWeb.com*, May 24, 2019a. As of August 1, 2019:
http://www.russianspaceweb.com/angara5.html

———, "Russian Space Program and Rocket Development in 2018," *RussianSpaceWeb.com*, August 6, 2019b. As of August 1, 2019:
http://www.russianspaceweb.com/2018.html

———, "The Soyuz Rocket Family," *RussianSpaceWeb.com*, October 22, 2019c. As of August 1, 2019:
http://www.russianspaceweb.com/soyuz_lv.html

Zelnio, Ryan, "A Short History of Export Control Policy," *Space Review*, 2006. As of August 1, 2019:
http://www.thespacereview.com/article/528/1